Lecture Notes in Earth Sciences 102

Editors:
S. Bhattacharji, Brooklyn
H. J. Neugebauer, Bonn
J. Reitner, Göttingen
K. Stüwe, Graz

Founding Editors:
G. M. Friedman, Brooklyn and Troy
A. Seilacher, Tübingen and Yale

Springer
Berlin
Heidelberg
New York
Hong Kong
London
Milan
Paris
Tokyo

Werner Smykatz-Kloss
Peter Felix-Henningsen (Eds.)

Palaeoecology of Quaternary Drylands

With 52 Illustrations, 2 in Colour

Editors

Professor Dr. Werner Smykatz-Kloss
Institut für Mineralogie und Geochemie
Universität Karlsruhe
76128 Karlsruhe, Germany

Professor Dr. Peter Felix-Henningsen
Institut für Bodenkunde und Bodenerhaltung
Heinrich-Buff-Ring 26
35392 Giessen, Germany

Cataloging-in-Publication Data applied for

Bibliographic information published by Die Deutsche Bibliothek.
Die Deutsche Bibliothek lists this publication in the Deutsche Nationalbibliographie; detailed bibliographic data is available in the Internet at <http://dnb.ddb.de>.

ISSN 0930-0317
ISBN 3-540-40345-0 Springer-Verlag Berlin Heidelberg New York

This work is subject to copyright. All rights are reserved, whether the whole or part of the material is concerned, specifically the rights of translation, reprinting, re-use of illustrations, recitation, broadcasting, reproduction on microfilms or in any other way, and storage in data banks. Duplication of this publication or parts thereof is permitted only under the provisions of the German Copyright Law of September 9, 1965, in its current version, and permission for use must always be obtained from Springer-Verlag. Violations are liable for prosecution under the German Copyright Law.

Springer-Verlag is a part of Springer Science+Business Media

springeronline.com

© Springer-Verlag Berlin Heidelberg 2004
Printed in Germany

The use of general descriptive names, registered names, trademarks, etc. in this publication does not imply, even in the absence of a specific statement, that such names are exempt from the relevant protective laws and regulations and therefore free for general use.

Cover design: Erich Kirchner, Heidelberg
Typesetting: Camera ready by author
Printed on acid-free paper 32/3142/du - 5 4 3 2 1 0

Contents

The Importance of Desert Margins as Indicators for Global Climatic Fluctuations (Introduction)

In various periods throughout the younger earth history comparable changes in climate occurred globally and simultaneously. Such global events can be reconstructed with the help of reliefs, sediments and palaeosoils and their specific morphological, chemical and mineralogical properties. Desert margins represent intersections between arid and humid ecosystems. Their geographical position will react very sensitively on climatic changes. The broad regions of transformation between recent humid ecosystems and the fully arid deserts are the proper areas where palaeoclimatically different phases will be remarkably recognized and interpreted.

Aeolian sediments, e.g. dunes, can be used as palaeoclimatic indicators: palaeodunes in todays more humid climate may indicate arid conditions at the time of their deposition. As an example, fossil dunes are widely distributed in the Sahel south of the Sahara. In resting periods of sedimentation a cover of vegetation appears, and chemical weathering processes and hence soil formation takes place on the sediments in humid climates (see Felix-Henningsen, Heine, Rögner et al., Smykatz-Kloss et al.). In deeper positions of the relief fluvial sediments in wadis and limnic sediments in palaeolakes and playas were deposited. They can be recognized by their sedimentary structures and by characteristic mineral associations, such as for example transformed evaporites (see Rögner et al., Schütt, Heine), by diatomites and lacustrine sediments (see Baumhauer et al.) or by bog ores (see Felix-Henningsen). At some rare occasions the coastline of a former lake is traced by fulgurites (see Sponholz). The organogenic components of soils and sediments mirror the palaeoecological conditions and changes (see Smykatz-Kloss et al.). The pollen communities in upper soils and sediments show the spectrum of the vegetation and thus deliver important criteria for palaeoclimates and relative ages (see Baumhauer et al.). Anthropogenic relicts in soils and sediments are a proof for humid phases. The existence of humid phases and their relative occurrences in the stratigraphical context and the kinds of sediments and palaeosoils allow the reconstruction of the frequency, relative age and character of palaeoclimatic changes (see Rögner et al., Mischke et al.). Absolute dating of aeolian sediments by using luminescence methods such al TL or OSL (see Jäkel, Smykatz-Kloss et al.) and organic substances (^{14}C) – where present – indicate the age position. If the sets of data are sufficiently dense, a picture can be obtained about the time periods of the humid and arid climate phases (Eitel et al., Rögner et al., Smykatz-Kloss et al.).

The signals of arid periods can partly be discovered widely distributed, e.g. over the desert margins to off-shore regions in the oceans. Thus, Leuschner, Sirocko et al. describe layers of (aeolian) dust from Saudi-Arabia in drilling profiles of the Arabian Sea: the geochemical and sedimentological evaluation of these palaeo-loesses in the marine sediment cores contributes to the reconstruction of palaeo-monsoon movements (Leuschner et al.).

Questions on the palaeo-ecological interpretation of drylands and desert margins are explored in the German working group *"desert margins"* and in many interdisciplinary projects. The group conferences are held annually in January at the Rauischholzhausen castle near Gießen. This working group, which has also acted as the German representation for several international geological correlation programmes (all concerned with desert research: IGCP 250, 349, 410), was established seven years ago by the editors of this volume. It is made up of approximately 50 geoscientists of (nearly) all disciplines: geomorphologists, geologists, mineralogists, geochemists, soil scientists, geochronologists, sedimentologists – as well as several palynologists, geobotanists and archaeologists.

At the beginning a pilot project built the core of the research (group) comprising nine projects from the edges of the Sahara (Reichelt, Baumhauer et al., Felix-Henningsen, Rögner et al., Schulz et al., Smykatz-Kloss et al., Sponholz) and of the Namib (Eitel et al., Heine). After a while the study areas were extended towards the north-west (Spain: Schütt, Günster) and – primarily – (north-) eastwards across the Arabic world (Leuschner, Sirocko et al.) towards Central Asia (Grunert & Lehmkuhl; Mischke, Hofmann et al.; Walther). Methodical questions on age analysis (dating of young sediments and aridic soils) and the correlation between chemical weathering (geochemistry, soil science) and palaeoecology are the themes that raise the regional and subject specific results onto a global scale (Jäkel; Eitel, Blümel & Hüser; Felix-Henningsen; Heine; Leuschner, Sirocko et al.; Rögner et al.; Schütt; Smykatz-Kloss et al.).

The investigation of the desert margins as suitable indicators for global climatic fluctuations belongs to the basic research in palaeoecology. The obtained results contribute to the efforts of several earth scientific disciplines in order to understand and reconstruct the causes, frequencies and time periods of palaeoclimatological events and changes. This is especially important on the background of the recent global temperature increase, which is mainly anthropogenetically initiated, and of regional climatic catastrophes. The prognosis of long-term consequences on the base of modelling exhibits many uncertainties concerning the frequency, duration and amplitude of natural climatic fluctuations.

Additionally, the results of studies on desert margins enrich our knowledge on the complexities of landscape formation and on the distribution pattern of their resources (e.g. soils and groundwater) in dependence on extremely different climatic conditions and changes. Ecosystems of savannahs and semi-deserts in the regions of desert margins and the people living there are endangered in their existence by short- and long-term climatic fluctuations. The research data of the working group contribute to a more pronounced understanding of these ecosystems: not only the studied structures and processes, but their development in time, their formation and disappearance under the influence of global climatic changes have to be regarded. Geomorphological research in these climatic regions (e.g. the desert margins) will only be effective if the various geo- and bioscientific disciplines will work together. The contributions in this volume may proof this. The realisation of these projects has been made possible due to the support of numerous friends, members of staff, helpers and organisations. We would like to thank all of them, mainly all the reviewers and correctors, very especially our experts of desert research and production, Simon Berkowicz (Jerusalem) and Wolfgang Klinke (Karlsruhe), and above all the "Deutsche Forschungsgemeinschaft" (German Research Foundation) for its generous financial support.

Karlsruhe and Gießen, May 2003

W. Smykatz-Kloss
(Karlsruhe)

P. Felix-Henningsen
(Gießen)

The chemistry of playa-lake-sediments as a tool for the reconstruction of Holocene environmental conditions - a case study from the central Ebro basin

Brigitta Schütt

Institute of Geographical Sciences
Free University of Berlin
Malteserstr. 74-100, D-12249 Berlin

Abstract

The focus of the presented study is the reconstruction of the Holocene limnic and drainage basin conditions of the Laguna de Jabonera, a today playa-lake-system in the Desierto de Calanda, central Ebro Basin, using the inorganic characters of the lacustrine sediments. Mineralogical fabric helped to reconstruct the overall geomorphic processes and gives clues to the synsedimentary limnic environment (paleosalinity). The chemical composition of the lacustrine sediments largely reflects the mineralogical composition, but the higher resolution of the geochemical data compared to the mineralogical data enables to stratigraphically split the extracted core profile into three stratigraphic units. Supplementally, it is demonstrated that statistics between chemical compounds point to the synsedimentary intensity of weathering and soil forming processes.

As for the lacustrine sediments investigated there are no data yet available a preliminary chronological framework is derived by comparison with results from neighbouring areas. Based on this the hypothesis is put forward that during the so-called *Little Ice Age* subhumid to dry-subhumid environmental conditions occurred. Also possibly during the *late Subboreal* distinct wetter environmental conditions than today prevailed. Additionally, it is demonstrated that in the most recent past human impact is causing increased erosion rates and, thus, increased deposition of detritals in the most recent lacustrine sediments.

1 Introduction

Core-based paleoenvironmental investigations of lacustrine settings have been mostly dominated by micro-paleontological and pollen-based studies. Inorganic features of lake sediments are predominantly studied using mineralogical analyses to characterize lake typology (depth of water column, lake phase, salinity). The focus of this research is the reconstruction of Holocene weathering conditions of a today playa-lake-system in the Desierto de Calanda, central Ebro Basin by the analyses of the chemical character of lacustrine sediments, supplementing information about the limnic environment derived from the mineralogical character.

2 Site description

The region of the Desierto de Calanda southwest the town of Alcañiz is characterized by a plain built of slightly cemented Miocene clay strata with paleochannels of calcareous sandstone (Riba et al., 1983). The receiving stream of the Desierto de Calanda is the Rio Guadalope, a tributary of the Rio Ebro.

The present climate of the area is subarid Mediterranean with mean annual precipitation between 300-350 mm. Precipitation peaks during autumn and spring when the region is under the influence of westerlies. Summer aridity lasts for four months. Present-day climatic conditions in the Desierto de Calanda induce a mean annual precipitation-evaporation-ratio (P/pET) of 0.45 (Garcia de Pedraza and Reija Garrido, 1994), that means, according to the classification of the aridity-indices as quoted by UNEP (1991), the Desierto de Calanda belongs to the semi-arid dryland regions. Under present conditions, the mean annual groundwater influx into the endorheic basins of the Desierto de Calanda amounts to c. 60 mm, the mean annual surface inflow comes to 15 mm (data estimated according to Sanchez Navarro et al., 1991).

In the Desierto de Calanda endorheic basins were formed by the combined processes of subsurface erosion of underlying gypsum layers and deflation of out cropping clay strata (Ibañez, 1973; Fig. 1). Thus, in an area of approximately 100 km^2 more than 20 endorheic basins varying in size were built (Sanchez Navarro et al., 1991, 23). The largest of these basins are several hundred metres in diameter and get periodically flooded. Only some basins, predominantly located in the northeast of the Desierto de Calanda, are completely desiccated (Ibañez, 1973). Paleochannels form the boundary of the endorheic basins and elevate up to 20 m above the lake floor. Present processes of surface erosion occur, but forms are peri-

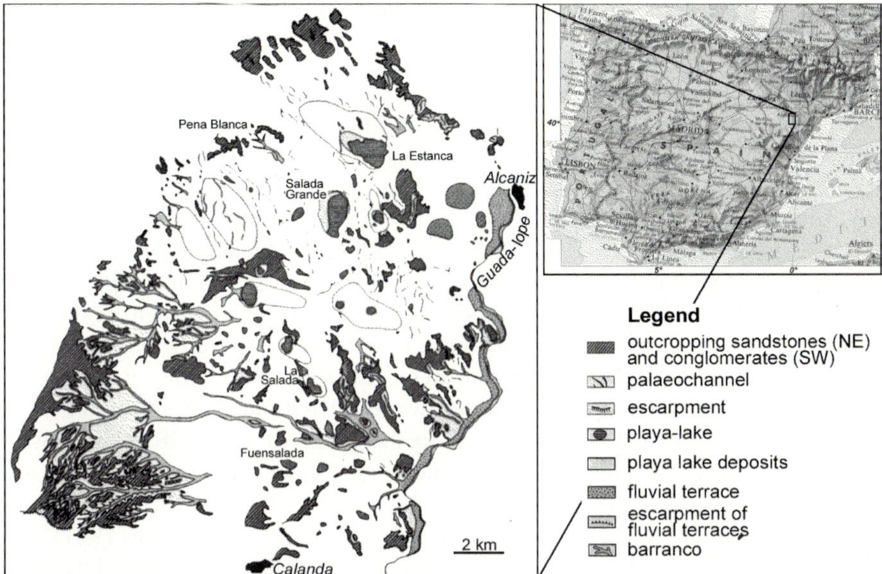

Fig. 1. Geomorphological-lithological map of the Desierto de Calanda (copy from Ibañez, 1973, 23)

odically removed by ploughing of the fields. The present solution of evaporitic rocks is reflected in high groundwater salinity (5765 mg/l TDS, n=13). Predominance of sulphates in the solutes (μ=3640 mg/l SO_4^{2-}, std.=2207, n=13) reflects their origin from solution of solid Miocene sulphates. Composition of cations also points to solution of sulphates with calcium, magnesium, and sodium ions predominating (μ_{Ca}=526 mg/l Ca^{2+}, std.=143; μ_{Mg}=457 mg/l Mg^{2+}, std.=305; μ_{Na}=569 mg/l Na^+, std.=730; n=13). During sampling in March 1994 table of perched groundwater varied between -0.8 m to -5.2 m below surface (all data are based on field data sampled from wells in the watershed of the Desierto de Calanda in March 1994).

Presently the area is being used for dry farming in which fallow land is used for pasture.

In the Desierto de Calanda cores were taken in the endorheic basins of the Salada Grande (easting 735 000, northing 4548 000; UTM coordinate system), the Laguna Pequeña (easting 733750, northing 4547 500; UTM coordinate system) and the Laguna de Jabonera (easting 736 600, northing 4547 500; UTM coordinate system). As sediments of all three endorheic basins point to the same sedimentary history, results shown in this paper are from the Laguna de Jabonera, an endorheic basin with a depth of c. 20 m and a lake floor diameter of c. 1 km. The lacustrine sediments discussed were extracted in the centre of the basin; coring went down to the Miocene bedrock in 310 cm depth below lake floor.

3 Methods

In general, cores were taken in the centre of the playa-lake-systems. To avoid core loss and to control sediment compaction by drilling, two parallel cores were taken with about 0.5 m vertical displacement. A modified Kullenberg corer with a hydraulic core catcher and a diameter of 40 mm was used to obtain undisturbed sediment samples.

Analyses of the sediments included first a sedimentological description to identify stratigraphical units by macroscopic characters. Preparation of samples started with drying them at 50° C in a drying cabinet and homogenizing them in an agate swing sledge mill. Organic and inorganic carbon contents were determined by an infrared cell in a LECO after burning in an O_2-flux (detection limit = 0.02 mass-% C). Analyses of mineralogical compounds were carried out by X-ray powder diffraction analyses using Cu K_{α}-radiation in the range of 2-70 °2θ with steps of 0.01 °2θ and each step measured for one minute. Concentrations of calcite and dolomite were estimated by calibrating of the intensity of major diffraction peaks of calcite and dolomite (cps) by inorganic carbon contents (Behbehani, 1987). The position of dolomite's major diffraction peak was determined to obtain data about the Mg-Ca-ratios of the carbonates after calibration of the diffractogram with reference to the major diffraction peak of quartz (Tennant and Berger, 1957; Langbein et al., 1981); data were traced with two decimals but are presented with three decimals wherever they show average values. Bulk chemistry of samples was determined by X-ray fluorescence analyses (Siemens SRS 2000). For interpretation

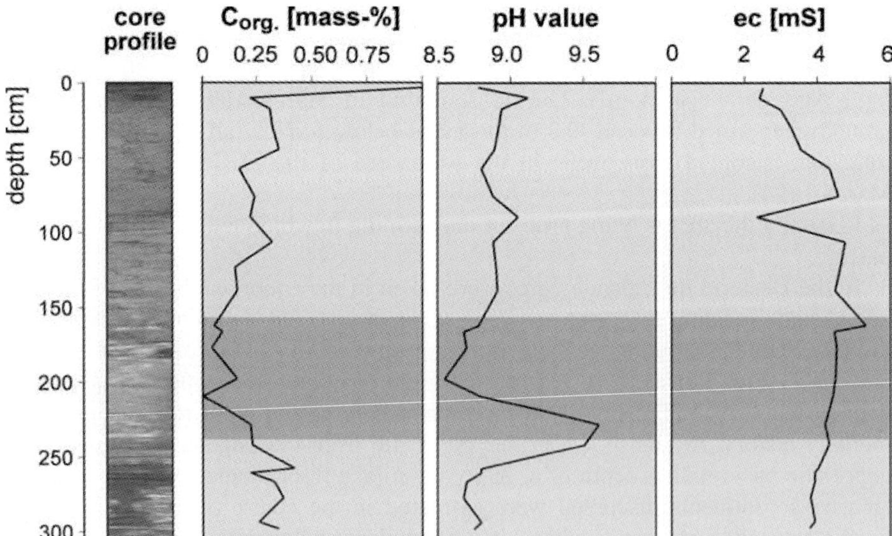

Fig. 2. Bedding of the core profile, organic carbon content (mass-% $C_{org.}$), pH-value, and electric conductivity (ec mS) of lacustrine sediments

and discussion of the lacustrine settings exposed sediments arc subdivided into stratigraphic units which are defined by their mineralogical and chemical composition.

Because of very low contents of organic carbon in the playa-lake-sediments, the technique of OSL-dating was chosen to obtain data, executed at the *Forschungsstelle Archäometrie* (Heidelberger Akademie der Wissenschaften, Max-Planck-Institut für Kernphysik). As data are not available yet correlation with investigations from neighbouring areas give a preliminary time scale.

Parallel to coring drainage basin characters were surveyed. Among the study of geological, geomorphological, and land use settings main emphasis was put on hydrological conditions. In the watershed level of perched-groundwater was measured in the wells using a light plumb line. Additionally, water samples from wells were taken and chemical composition was analysed in the laboratory using ion chromatography.

4 Sediment character

4.1 General sedimentary fabric

Lacustrine sediments from the Laguna de Jabonera are of a greyish brown at the basal layers (7.5 YR 4/4) and to the top continuously change to a more reddish colour (2.5 YR 5/4 in 240 cm depth), repeatedly interstratified by fibrous gypsum. From 240 cm depth to the surface sediments are uniformly brownish grey. The organic carbon contents in the sediments reach 1.02 mass-% $C_{org.}$ close to the lake-bed surface but decrease rapidly below 5 cm depth only to oscillate around μ_{Corg}=0.19 mass-% $C_{org.}$ (std.= 0.153, n=24). The sediments are slightly basic (μ_{pH}=8.9, $std._{ph}$= 0.23, n=25), only between 230 and 250 cm depth the pH rises to 9.6 (Fig. 2).

4.2 Mineralogical composition

The whole core is characterized by the simultaneous occurrence of quartz, gypsum, and calcite with gypsum predominant in the parts below 130 cm depth, and carbonates in the upper part (Table 1). Thin sections show that along the whole core the mineralogical composition is idiomorphic carbonates and gypsum embedded in an alternating medium- to fine-grained groundmass of carbonates and gypsum; only in the most recent sediments carbonates are detrital. Dehydrated sulphates (anhydrite) can be detected along the entire core profile as traces. Other than sulphates halites exist as evaporitic minerals. Their concentration decreases to the top. They mainly appear as a minor component in the core section below

Table 1. Mineralogical composition of lacustrine sediments

stratigraphic unit	depth [cm]	*Carbonates* Calcite	Dolomite	*Sulphates* Gypsum	Anhydrite	Halite	Quartz	Phyllo-silicates	Goethite
	3	+++	++	++	+	++	+++	++	+
	10	+++	++	+	+	+	+++	++	+
	18	+++	++	++	+	+	+++	++	
	29	+++	++	++	+	+	+++	++	+
	45	+++	++	++	+	+	+++	++	
3	58	+++	++	+++	+	+	++	++	
	77	+++	++	++	+	+	+++	++	+
	91	+++	++	++	+	+	+++	++	+
	108	+++	++	+++	+	+	+++	++	
	125	+++	++	++	+	+	+++	++	
	141	++	++	+++	+	+	+++	+	+
	165	+	+++	++	+	+	+++	++	
	169	+	++	+++	+	++	++	+	
	180	++	+++	+++	+	++	++	++	
2	201	++	++	+++	+	++	++	++	
	213	+	++	+++	+	+	+	+	
	232	++	++	+++	+	++	++	+	
	246	+	+++	+++	+	+	+++	+	+
	261	++	++	+++		++	++	+	
	265	++	++	+++	+	++	++	+	
1	271	+	++	+++	+	++	+++	+	
	282	+	++	+++		++	+++	+	+
	292	++	++	+++	++	++	++	+	
	303	+	++	+++	+	+	+++	+	

max.counts p.s. +++ major components
++ minor components + traces

165 cm depth (stratigraphic units 1 and 2) and only as traces above (stratigraphic unit 3). Calcite contents continuously increase from bottom to top ($\alpha<0.001$) while above 165 cm depth dolomite contents decrease continuously ($\alpha<0.05$).

The dolomite contents alternate strongly below 165 cm depth (Fig. 3). Between 180 and 220 cm depth the average dolomite content amounts to 8.8 mass-%, in 246 cm depth its content rises up to 18.3 mass-% and fluctuates below 246 cm depth between 4.6-13.6 mass-%. Dolomite's major diffraction line varies along the whole core profile about μ_{dol100}=30.828 °2θ Cuk_{α} (std.= 0.046, n=25). In stratigraphic unit 2 the position of dolomite's major diffraction line averages $\mu_{dol100;2}$=30.859 °2θ Cuk_{α} (std.= 0.042, n=6) and, thus, does not differ significantly from the angles of diffraction in the underlying and overlying sediments ($\alpha>0.05$).

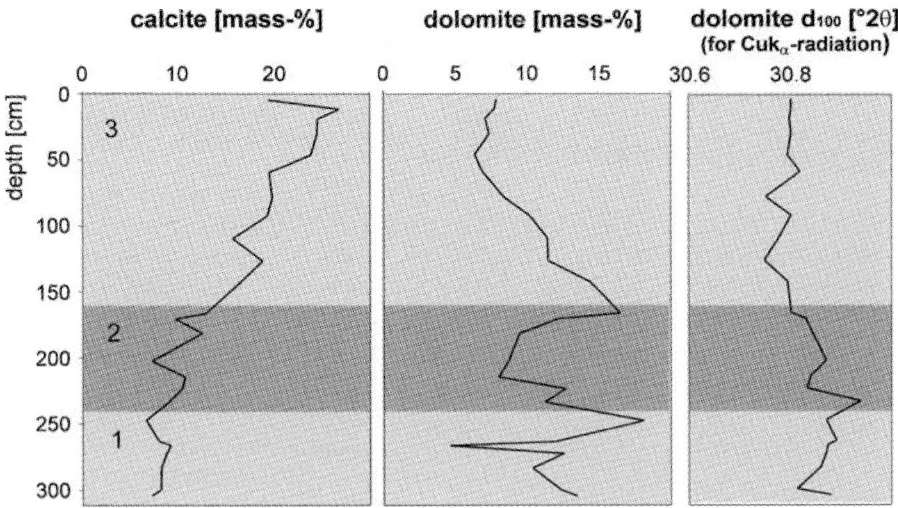

Fig. 3. Calcite and dolomite contents (mass-%) of lacustrine sediments and location of the diffraction angle of dolomite's major diffraction peak 100 (dol_{100} °2Θ Cu K_{α}-radiation)

4.3 Chemical composition

4.3.1 Major elements

As lacustrine sediments of the Laguna de Jabonera are subdivided into three stratigraphic units (cf., chapter 4.4) mean chemical composition of each stratigraphic unit is summarized in Table 2. These data show evidently that in stratigraphic unit 1 concentrations of SiO_2, MnO, and TiO_2 are distinctly higher and concentrations of S (sulphur content expressed as SO_3) and CaO distinctly lower than in the overlying stratigraphic unit 2 (166-240 cm depth; Fig. 4). At a depth of 165 cm (stratigraphic unit 3; 0-165 cm depth) there is again a marked change in the chemical sediment composition. From the lower to the higher layer, calcium oxide and sulphur concentrations decrease distinctly while the concentrations of the most metal oxides and of silica increase (Table 2). In 50 to 70 cm depth the sulphur graph again shows a marked deflection, coinciding for a short time concentrations of silica and metal oxides are reduced (Fig. 4).

Towards the top continuously increasing phoshate contents ($\alpha<0.05$) are correlated to the sediment's calcite contents:

$$P_2O_5 \text{ mass-\%} = 0.06237 + 0.00386 * \text{calcite mass-\%} \qquad \text{(eq. 1)}$$

$n=25$, $r=0.87392$, adj. $r^2=0.75346$

Table 2. Laguna de Jabonera/I: Average chemical composition of core sections 1-3 and t-test matrix

		section 1 [-310;-241] cm depth n=7	section 2 [-240;-166] cm depth n=6	section 3 [-165;0] cm depth n=12	probability level of t-test section 1 vs. section 2	 section 2 vs. section 3
SO_3 [mass-%]	µ	16.76	23.02	5.02	α<0.01	α<0.001
	std.	3.522	3.109	2.606		
SiO_2 [mass-%]	µ	36.95	24.09	40.69	α<0.001	α<0.001
	std.	4.076	2.851	2.21		
TiO_2 [mass-%]	µ	0.42	0.328	0.56	α<0.01	α<0.001
	std.	0.040	0.041	0.030		
Al_2O_3 [mass-%]	µ	8.47	7.93	14.28	α>0.05	α<0.001
	std.	0.358	1.126	0.948		
Fe_2O_3 [mass-%]	µ	2.99	2.80	5.17	α>0.05	α<0.001
	std.	0.17	0.407	0.390		
MnO [mass-%]	µ	0.047	0.04	0.067	α<0.05	α<0.001
	std.	0.0049	0.006	0.0107		
MgO [mass-%]	µ	7.95	8.60	7.25	α>0.05	α<0.05
	std.	0.587	0.902	1.599		
CaO [mass-%]	µ	23.53	30.14	22.43	α<0.01	α<0.001
	std.	3.009	2.418	1.783		
Na_2O [mass-%]	µ	1.16	1.36	1.75	α<0.05	α<0.001
	std.	0.126	0.147	0.347		
K_2O [mass-%]	µ	1.62	1.58	2.64	α>0.05	α<0.001
	std.	0.078	0.290	0.267		
P_2O_5 [mass-%]	µ	0.090	0.097	0.144	α<0.05	α<0.001
	std.	0.0058	0.0052	0.0162		

But stratigraphic units are not only reflected in the average chemical composition, also the quality of the statistics of chemical parameters points to the subdivision of the lacustrine sediments into the stratigraphic units defined (cf. chapter 4.4). All over the lacustrine sediments only graphs of SiO_2 and TiO_2 are correlated positively to each other and negatively to the graph of SO_3 ($\alpha<0.01$). In contrast, statistics between the other chemical compounds are highly variable. In stratigraphic unit 1, the statistics between the contents of the various elements are collectively weak and only prove to be statistically significant in a few cases ($\alpha<>0.05$; Appendix Table A1). In stratigraphic units 2 and 3 components originated predomi-

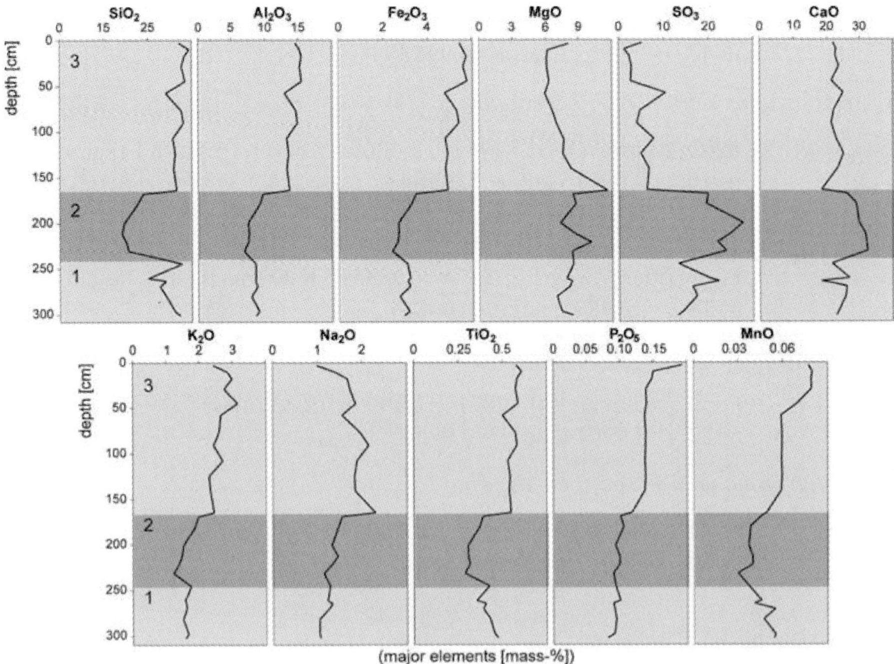

Fig. 4. Chemical composition of lacustrine sediments (major elements mass-%)

nantly from clay strata as SiO_2, TiO_2, Al_2O_3, K_2O, and Fe_2O_3 are, in general, correlated positively to eachother ($\alpha<0.05$). Simultaneously, in stratigraphic units 2 and 3 the concentrations of SiO_2, TiO_2, Al_2O_3, and Fe_2O_3 correlate negatively with the sulphur contents ($\alpha<0.05$). In stratigraphic unit 2 calcium oxide contents are generally subject to a positive trend to sulphur contents while a negative trend with silica and most of the metal oxides occurs ($\alpha<0.01$). In stratigraphic units 1 and 3 statistics of calcium oxide to the other components are, in general, weak. Summarizing, it can be emphasized that statistics between chemical substances are weak or entirely non existent in stratigraphic unit 1 ($\alpha<>0.05$) while in stratigraphic units 2 and 3 statistics between various component contents are highly significant, but with distinct higher significance levels in stratigraphic unit 2 than in stratigraphic unit 3 (see appendix Table A1).

In addition, for SiO_2, TiO_2, Al_2O_3, K_2O, and Fe_2O_3 the slope of the regression line was tested. For linear regressions with the less soluble component as the independent variable, e.g. as $Al_2O_3=f\{Fe_2O_3\}$, the slope is regularly steepest in stratigraphic unit 2 ($\alpha<0.01$):

stratigraphic unit 1 (310;241 cm depth):

$$Al_2O_3\ (\text{mass-\%}) = 0.23691 + 1.74769 * Fe_2O_3\ (\text{mass-\%}) \quad \text{(eq. 2)}$$

$n=18$, $r=0.82951$, adj.$r^2=0.62570$

stratigraphic unit 2 (240;166 cm depth):

$$Al_2O_3 \text{ (mass-\%)} = 0.19122 + 2.76116 * Fe_2O_3 \text{ (mass-\%)} \quad \text{(eq. 3)}$$
n=12, r=0.99822, adj.r^2=0.99555

stratigraphic unit 3 (165;0 cm depth):

$$Al_2O_3 \text{ (mass-\%)} = 1.84947 + 2.40694 * Fe_2O_3 \text{ (mass-\%)} \quad \text{(eq. 4)}$$
n=10, r=0.99015, adj.r^2=0.97843

stratigraphic unit 1 (310;241 cm depth):

$$Fe_2O_3 \text{ (mass-\%)} = 1.23577 + 1.08359 * K_2O \text{ (mass-\%)} \quad \text{(eq. 5)}$$
n=18, r=0.49884, adj.r^2=0.09861

stratigraphic unit 2 (240;166 cm depth):

$$Fe_2O_3 \text{ (mass-\%)} = 0.59871 + 1.39093 * K_2O \text{ (mass-\%)} \quad \text{(eq. 6)}$$
n=12, r=0.99014, adj.r^2=0.97547

stratigraphic unit 3 (165;0 cm depth):

$$Fe_2O_3 \text{ (mass-\%)} = 2.60917 + 0.96935 * K_2O \text{ (mass-\%)} \quad \text{(eq. 7)}$$
n=10, r=0.66275, adj.r^2=0.38315

4.3.2 Traces

Considering the heavy-metal contents of the lacustrine sediments, evidence of lead cannot be found throughout the core section. The copper, chrome, nickel, and zinc content variations within the lacustrine sediments correspond with the metal oxide contents (Al_2O_3, Fe_2O_3, TiO_2; cf. Fig. 4) and reach maximal values in stratigraphic unit 3 with μ_{Cr}=61 ppm Cr (std.$_{Cr}$=14.5), μ_{Cu}=94 ppm Cu (std.$_{Cu}$=4.2), μ_{Ni}=38 ppm Ni (std.$_{Ni}$= 13.8) und μ_{Zn}=81 ppm Zn (std.$_{Zn}$=7.3) ($\alpha<0.001$). Opposed to this, in stratigraphic units 2 and 3 strontium concentrations correlate highly linear with sulphur contents ($\alpha<0.001$). Average strontium concentrations prove to have lower values in stratigraphic unit 3 with μ_{Sr}=550 ppm Sr (std.$_{Sr}$=175) than in stratigraphic unit 2 (μ_{Sr}=2830, std.$_{Sr}$=851, n=6) ($\alpha<0.001$); concurrently, the average strontium contents in stratigraphic unit 2 are lower than in stratigraphic unit 1 (μ_{Sr}=4548, std.$_{Sr}$=3120, n=7) ($\alpha<0.01$).

4.4 Defining stratigraphical units

Differentiation of the core profile into the three stratigraphic units

- 1 (241 to 310 cm depth),
- 2 (166 to 240 cm depth),
- 3 (0 to 165 cm depth)

was proven by discriminant analysis using the variables '*allothigenic components*' and '*position of dolomite's major diffraction peak*' (dol_{100}). The bulk of allothigenic minerals corresponds to detritals illuviated from the watershed where Miocene clay strata build predominantly the outcropping bedrock. Thus, '*allothigenic components*' are substituted by the sum of SiO_2, Al_2O_3, Fe_2O_3, and MnO (mass-%). As the '*authigenic components*' are the counterpart of the '*allothigenic components*' and both sum up to 100 mass-%, these variables are highly correlated ($\alpha<0.001$). Thus, the variable '*authigenic components*' is not suitable as the second variable for discriminant analysis. As the second variable for discriminant analysis the position (°2θ) of dolomite's major diffraction peak 'dol_{100}' was choosen, as this variable is only weakly correlated to the '*allothigenic components*' ($\alpha>0.05$) and, at the same time, the position of dolomite's major diffraction peak reflects the limnic conditions during formation of '*authigenic components*' - and, therefore, represents the other fraction of the lacustrine sediments.

Distribution of data lead to a differentiation into three stratigraphic units corresponding to the clusters shown in Fig. 5. Discriminant function $Y_{1/2}$

$$Y_{1/2} = 1.084*\text{allothigenic minerals (mass-\%)} + 0.477*dol_{100}\ (°2\theta\ Cuk_{\alpha}) \quad \text{(eq. 8)}$$

allows in all cases correct assignment of data to stratigraphical units 1 and 2 (n=13; Wilk's L=0.205). Also discriminant function $Y_{2/3}$

$$Y_{2/3} = 0.956*\text{allothigenic minerals (mass-\%)} - 0.170*dol_{100}\ (°2\theta\ Cuk_{\alpha}) \quad \text{(eq. 9)}$$

allows correct assignment of data to stratigraphical units 2 and 3 in all cases (n=18; Wilk's L=0.078).

5 Paleoenvironmental indications by the inorganic character of lacustrine sediments from playa-lake-systems

5.1 Information about paleolimnic environment from mineralogy

The mineralogical composition of the detrital lacustrine sediments in the area investigated is determined by the parent material. Apart from allochthonously deposited detritus authigenic carbonates and sulphates make up the mineralogical setting of the lacustrine sediments. Early diagenetic processes, steered by salinity and chemistry of brines, i.e. pore water, can modify the mineralogical properties. The influence of brine salinity on the authigenic mineral fabric and early diagenetic modification of minerals is predominant. Therefore, the mineralogical sediment properties provide some valuable information for the reconstruction of the paleoenvironment.

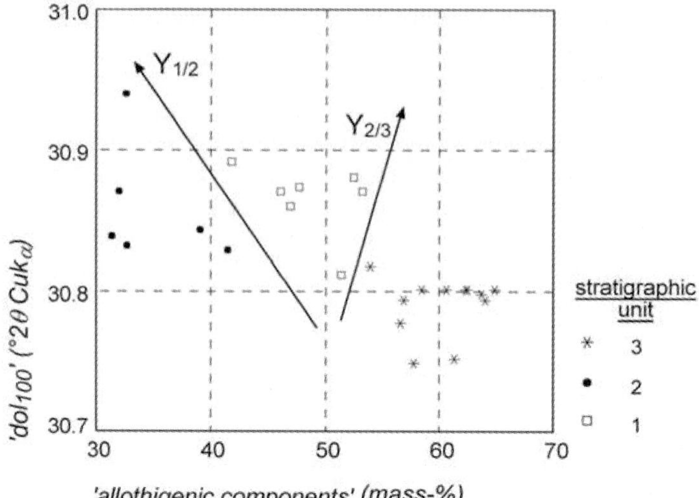

Fig. 5. Scatter-plot of sediment's dolomite's major diffraction peak '*dol_{100}*' and '*allothigenic components*', substituted by the sum of SiO_2, Al_2O_3, Fe_2O_3, and MnO (mass-%) seperated for sections 1-3

For the analysis of the mineralogical composition X-ray powder diffraction of the samples was chosen as this method is a basic tool in the mineralogical analysis of sediments. Data resulting from X-ray powder diffraction are only semiquantitative as shown in Table 1, but with regard to the chemical sediment character also allow to generate discrete mineral contents (cf. Fig. 3). Additionally, this method offers as a by-product information about carbonate's lattice distances and, thus, information about Mg-Ca-ratios of carbonates as presented in chapter 4.2. Thus, even if for detection of carbonate's Mg-Ca-ratios thermal methods such as DTA and DSC provide a higher accuracy, the position of dolomite's major diffraction peak gives an idea on carbonate's Mg-Ca-ratio in the sample. Beneath step-width and goniometer's scan velocity determination accuracy of the position of carbonates major diffraction peak depends on the cristallinity of the the sample and sample's grain-size distribution.

5.1.1 Mineral fabric

Quartz, calcite, and gypsum make up the mineralogical major components of the lacustrine sediments of the Laguna de Jabonera. The analysis of these major components within the lacustrine sediments yields information on some conspicuous environmental changes (Fig. 6).

During periods with a negative water balance a sparse vegetation cover within the catchment, triggering aeolian processes and overland flow, leads to transposition of the parent material due to ephemeral to periodic heavy rains (Dunne et al., 1991; Rogers and Schumm, 1991). Lacustrine sediments which suit these processes in the Laguna de Jabonera basin include detrital carbonates, quartz, and silicates. Dry environmental conditions also mean a reduction of groundwater recharge and, corresponding, reduced subsurface flow (Horton, 1945). In the Desierto de Calanda surface input from evaporitic sediments may be excluded, as there are almost no outcrops of evaporitic sediments within the watershed. Consequently, lacustrine settings deposited under arid to subarid environmental conditions are rich in allochthonous components (clay minerals, detrital carbonates, and quartz), while autochthonous sediments (carbonates, sulphates, halites) are only of secondary importance. Vice versa, wetter periods allow a denser vegetation cover so the topsoil is protected from being eroded by wind or water. Low relief and vegetation cover support infiltration and, therefore, increase subsurface flow (Horton, 1945; Morisawa, 1959). Precipitation of evaporites in the Laguna de Jabonera therefore has to be explained by groundwater increase and, thus, increased subsurface inflow of aqueous solution, which will mostly take place in response to more humid conditions (Schütt, 1998a, 1998c). Consequently, during wetter conditions in the Laguna de Jabonera basin the reduced erosion combined with concurrent higher influx of sulphate-rich groundwater (SO_4^{2-}-Ca^{2+}-Mg^{2+}-Cl^--

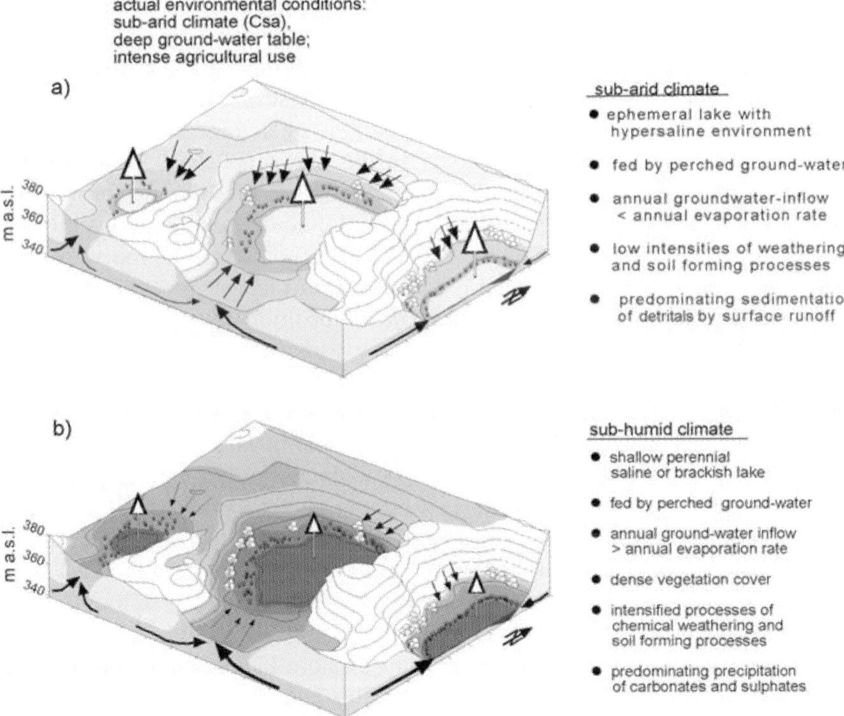

Fig. 6. Reconstruction of climatic influence of environmental and depositional conditions of the Desierto de Calanda: (a) sub-arid climate and (b) sub-humid climate

type) results in increased evaporite concentrations in the correlating lacustrine settings.

Taking these facts into account, the increased sulphate contents in the sediments of core section 2 trace back to a phase of well-balanced water-budget conditions with raised groundwater inflow and thereby high inflow rates of sulfatic aqueous solutions. The detrital carbonates and quartz in the sediments increase to the top because of worsening water-balance conditions and the consequently reduced inflow of aqueous solutions with the groundwater coinciding with increased erosion. A growing influence of soil erosion must also be taken into consideration for the formation of these most recent sediments (Faust and Diaz del Olmo, 1997; Stevenson et al., 1991). These predominant semiarid to dry-subhumid climatic conditions during the sedimentation of the uppermost strata (stratigraphic unit 3) were only briefly interrupted, indicated by a short-term increase of the sulphate contents in the sediments between 50 to 60 cm depth. A similar but less distinct deflection of sulphur graph occurs in stratigraphic unit 1 between 260 to 270 cm depth and, thus, also points to short-term wetter environmental conditions during an altogether dry-subhumid to subarid climatic period during deposition of stratigraphic unit 1.

5.1.2 Carbonates and phosphates

Calcareous mud is the predominant form of carbonate deposits in the playa-lake-system investigated. Its formation can be explained by strong mechanical stress on detrital carbonates (Kelts and Hsü, 1978) as well as by authigenic calcite precipitation (Schröder et al., 1983). As salinity in the playa-type lake investigated is high, and as magnesium as an indispensable prerequisite of dolomite formation is available (Folk and Land, 1975), also authigenic dolomite precipitation is possible (van der Borch, 1976; van der Borch and Lock, 1979). Essentially, the autochthonous development of calcareous mud (=automicrite) in lacustrine environments is due to water chemistry changes effected by decomposition of organic matter, biological assimilation of CO_2, or temperature increases with consequential salinity deviations (Flügel, 1978).

In general, the Mg-Ca-ratio of the dolomites turns out to be a helpful indicator for the reconstruction of paleosalinity. If $d_{10\underline{1}4}$ dolomites show lattice spacing higher than 2.8879 Å, the dolomite is non-stoichiometric and called 'protodolomite'. As the radii of calcium-ions (R_{Ca2+}=1.00) are bigger than the radii of magnesium-ions (R_{Mg2+}=0.72) the lattice spacing of dolomite becomes larger the smaller the Mg-Ca-ratio of dolomite is (Langbein et al., 1981). This is portrayed when using an X-ray-diffractogram, where the major diffraction peak of dolomite shifts from 30.94 °2θ Cuk$_\alpha$ to smaller angles with decreasing Mg-Ca-ratio (Royse et al., 1971; Tennant and Berger, 1957). Because magnesium-ions relatively accumulate whilst brines are confined and calcium carbonates and sulphates are precipitated (Morrow, 1979), the Mg-Ca-ratio of the dolomites reflects lake water salinity during dolomite formation, regardless whether dolomites originate from authigenic or early diagenetic processes (Folk and Land, 1975; Morrow, 1982).

But at all, paleoenvironmental conditions in the watershed of the Laguna de Jabonera as reconstructed on the base of the overall mineral fabric are not reflected in significant changes of lake water salinity as shown by the Mg-Ca-ratio of the dolomites because in the sediments of the Laguna de Jabonera changes in position of dolomite's major diffraction peak do not vary significantly between stratigraphic units (Fig. 3). High inflow rates of solutes during wetter phases get compensated by high evaporation rates during drier phases. Thus, the salinity of lake water is always high and without marked changes. Consequently, for the lacustrine sediments of the Laguna de Jabonera the Mg-Ca-ratio of dolomites does not offer useful information for the reconstruction of paleohydrology. In stratigraphic unit 3 the negative trend between calcite and dolomite, corresponding to from bottom to top increasing calcite contents and decreasing dolomite contents, might be due to the downward continuing process of early diagenetic dolomitization. As Mg-Ca-ratios of dolomite show no significant changes in these stratigraphic unit it can be excluded that trends of calcite and dolomite concentrations with depth are due to changes in salinity and, thus, to changing conditions of authigenic carbonate formation.

During precipitation of authigenic calcite dissolved phosphor can be combined and precipitated as apatite ($Ca_5(PO4)_3OH$ or $Ca_5(PO_4,CO_3,OH)_3(F,OH)$) (Stumm and Leckie, 1971; Müller, 1997). In addition, dissolved phosphor compounds can be bounded to iron and aluminium (Eckert et al., 1997). In contrast, during early diagenetic dolomitization magnesium can substitute the PO_4-ions and sodium as well as strontium can substitute phosphate in a hypersaline environment (Nathan, 1984). Nevertheless, the linear, close connection between phosphor(V)-oxide and calcite contents in the lacustrine sediments of the Laguna de Jabonera suggests the importance of calcite for the phosphor precipitation in these sediments. The phosphor(V)-oxide contents in the lacustrine sediments are then a result of primary apatite formation and secondary displacement of phosphor by sodium, magnesium or strontium ions. Due to an altogether low phosphor(V)-oxide content in the sediment (μ_{P2O5}=0.12 mass-% $P_2O_5 \cong 0.28$ mass-% apatite) it is not possible to detect apatite in the sediments of the Laguna de Jabonera by X-ray powder diffraction.

5.1.3 Sulphates

Alongside gypsum ($CaSO_4 \bullet 2H_2O$), also anhydrite ($CaSO_4$) was found as traces in the samples of the lacustrine sediments of the Laguna de Jabonera. In general, gypsum gets precipitated autochthonously from brines with a density >1.115 g/cm³ (Sonnenfeld, 1984, 102). In contrast, anhydrite can be precipitated autochthonously or might be a diagenesis product from gypsum (Usdowski, 1974). Diagenesis of gypsum is controlled by temperature, pressure, and lake water salinity (Holser, 1979). Anhydrites originating from dehydration of gypsum, in general, show strontium concentrations around 2200 ppm Sr corresponding to the average strontium content of gypsum (Usdowski, 1973). In contrast, in brines strontium gets relativley enriched with continuing evaporation; thus, autochthonous anhydrites contain 20000 to 30000 ppm Sr (Usdowski, 1974). Strontium concentrations in the lacustrine sediments of the Laguna de Jabonera, in general, range in the dimension 10^3 and, thus, point to anhydrates formed by dehydration of gypsum – that means by diagenetic processes. Altogether, significant higher strontium concentrations in sediments of stratigraphic unit 1 indicate a distinct negative water balance during gypsum precipitation. In contrast, low strontium concentrations in the most recent sediments reflect 'dilution'-processes by high input rates of detritals (quartz, silicates, carbonates) in consequence of soil erosion processes (Stevenson et al., 1991; cf. chapter 5.2.3).

5.2 Information about erosion and weathering conditions from sediment chemistry

Analogously to the bulk chemistry of the lacustrine sediments the catchment properties of the lakes are recorded and analyzed. That means if occasion arises, the petrography, structure, and relief of the catchment, as well as hydrographical und hydrogeological catchment properties were investigated, as these characters can limit the applicabilty of individual geochemical sediment properties as paleoenvironment indicators. In opposition to this, climatic factors influence the intensity of the material-mobilisation processes. For the chemical weathering this results from the availability of H_2O as the reaction agent and from the direct influence of the temperature to the speed of the reaction. The influence of the climate on the biomass production and consequently on the provision of vegetable matter should also be considered. Humic acids are produced when vegetable matter is decomposed; they influence the soil pH and, thereby, modify intensity of weathering processes. The climate can also have an effect on the erosion processes which can interfere with weathering and soil-formation processes.

5.2.1 Organic Carbon

The organic carbon is subject to early diagenetic decomposition processes, both, in the limnic environment as well as in the lacustrine sediments (Meyers and Ishiwatari, 1993). These processes take place during different redox-conditions (Evans and Kirkland, 1988) and, as a function of time, can lead to the complete consumption of the organic carbon (Lerman, 1979). The organic carbon in the lacustrine sediments of the Laguna de Jabonera is allochthonous and was transported from the catchment into the basin as a result of surface erosion, or it is autochthonous and developed owing to limnic biomass production (Håkanson and Jansson, 1983). The decompositon processes of organic matter are accelerated in aerobic environments with the presence of light and high temperatures (Vallentyne 1962). Accordingly, organic carbon in sediments close to the surface in ephemeral lakes is subject to quickened decomposition. Hence, the known high primary production in saline environments (Evans and Kirkland, 1988) does not consequently effect increased organic-carbon contents in the sediment. The graph of organic carbon concentration as shown in Fig. 2 essentially reflects the early diagenetic decomposition of organic matter advancing from the bottom to the top.

Because the components of phytoplankton are generally more soluble than allochthonous plant deritus (cellulose, chitin, lignin) (Vallentyne, 1962) the increased organic carbon contents in stratigraphic unit 1 possibly suggest intensified deposits of difficultly decomposable plant deritus. Additionally, the increased organic carbon concentrations can be an indicator for an altogether increased accumulation rate which induces a relative decrease of the organic carbon decomposition rate in the sediment (Lerman, 1979, 392). Both processes, nevertheless, likewise indicate intensified erosion dynamics for the Laguna de Jabonera catchment during sediment deposition of stratigraphic unit 1 and, thus, confirm subarid to dry-subhumid environmental conditions as already reconstructed by the overall mineral fabric.

5.2.2 Mean chemical composition

The mean chemical composition of the lacustrine sediments largely reflects the mineralogical composition. However, the higher resolution of the geochemical data compared to the mineralogical data enables to stratigraphically split the extracted core profile into three stratigraphic units (eq. 8, eq. 9). Correspondingly, not only the ascendency of gypsum in the sediments of stratigraphic unit 2 can be verified by the graphs of the sulphur-trioxide concentration in the sediment, moreover, by comparing stratigraphic units 1 and 2 an increased importance of sulphatic precipitations in unit 2 can be observed. Here, the mean silica-, titanium-, aluminium-, iron-, and manganese-oxide contents show exact opposite proportional values to these highest sulphur-trioxide contents by having their lowest values in stratigraphic unit 2.

One can deduce from the conclusions in the chapter on the mineralogical composition that the increased concentrations of authigenic minerals (sulphates, authigenic calcites) in the lacustrine sediments result from multiplied inflow of aqueous solution via groundwater, whereas increased detrital contents (detrital carbonates, silicates, quartz) in the lacustrine sediment reflect relatively intensified erosion dynamics in the catchment. Therefore, the raised mean sulphur-trioxide contents in stratigraphic unit 2 conclude that the sediments of this unit were deposited in a wetter phase. Correspondingly, the sediments affected more by detritus in stratigraphic units 1 and 3 were deposited in an environment controlled by arid conditions (cf. chapter 5.1.1). The comparison of the mean chemical composition of the sediments of stratigraphic unit 3 with that of stratigraphic unit 1 indicates the greater importance of authigenic mineral formations (SO_3 factor 3) and lower importance of detrital minerals (SiO_2 factor 0.9) in stratigraphic unit 1. In this valuation one must also consider the growing land-use intensity with time which has led to magnified erosion processes ('soil erosion') in the most recent past (Stevenson et al. 1991; Faust, 1995). In the sediments of stratigraphic unit 3 these soil erosion processes cause an increase of the detrital components. Moreover, the increased gypsum resp. sulphur-trioxide contents in 50 to 60 cm depth and 260 to 270 cm depth point to a short phase of equable water balance with increased groundwater inflow and, thereupon, proceeding increased rates of evaporite precipitation.

A syn- or postsedimentary change of the sediment chemistry by solution and displacement of silica containing sediments can largely be ruled out for lacustrine sediments with mean pH values of μ_{pH}=8.9 (Fig. 2) because, in this environment, the solubility of silica and aluminium is only slightly higher than in a neutral environment (SiO_2 factor 1.5). The solubility of these substances only increases exponentially when basicity goes up (SiO_2 solubility at pH 9.5 factor 2.6, at pH 10 factor 5) (Krauskopf, 1956; Degens, 1962).

5.2.3 Statistics between the contents of selected elements

Chemical weathering and soil forming processes depend on the availability of water and both increase in intensity with growing precipitation-evaporation-ratios (Krauskopf, 1967; Chesworth, 1992). The resulting differentiation of the soil profile increases with the intensity of these processes. During phases of well-balanced water-budget the erosion rates are low due to denser vegetation (Rogers and Schumm, 1991). Regarding only gradual progress of erosion processes, the correlated lacustrine sediments show good statistics between the individual chemical substances owing to the differentiated soil profiles induced by chemical weathering and soil-formation processes in a humid to subhumid environment (Krauskopf, 1967).

Trend of statistics

In the three stratigraphic units defined the connections between the sulphur trioxide contents and the silica and titanium oxide contents are subject to a negative trend (α<0.001) as well as negative trends between aluminium and iron oxide contents to sulphur trioxide can be noticed in stratigraphic units 2 and 3 (α<0.01). This indicates the different process structures which led to the deposition of these substances. While the sulphur was dissolved by groundwater and precipitated, due to evaporation, in the Laguna de Jabonera basin after being transported there, the silica, titanium, aluminium and iron in the lacustrine sediments, predominantly detritically linked, were washed into the basin as a result of surface erosion in the catchment (Sanchez Navarro et al., 1991).

Quality of statistics

In stratigraphic unit 2 the concentrations of the silica, titanium, aluminium, iron, manganese, potassium, and sodium oxide contents are positively correlated with eachother (α<0.01) which points towards the common origin of these substances from the weathering and erosion of the outcropping Miocene claystrata and Pliocene paleochannel fillings in the catchments. Comparatively, in stratigraphic units 1 and 3 the graphs of the silica, titanium, aluminium and iron oxide contents are also positively correlated (α<0.05), yet, further statistically significant connections with other component contents in the sediment are missing. Under semiarid conditions the weathering and soil-formation processes, and with that the chemical soil-profile differentiation, are limited (Krauskopf, 1967, 81). The simultaneously high erosion rates (Dunne et al., 1991) result in the correlated lacustrine sediments in 'random' chemical properties, that means statistics are incidental and are not subject to any regularities. The quality of statistics between the contents of different chemical substances is influenced by the mobility of the individual chemical parameters. The decreasing ionic potential of a substance makes the quality of its statistics with the concentrations of more stable substances go down. In addition, in stratigraphic unit 3, an intensified influence of phyllosilicates, occurring here as a minor component, particularly on the varying K_2O- and Na_2O contents must be taken into consideration due to the tying of the K^+-ions into the crystal lattice or ionic absorption (Heim, 1990).

Trend of regression line

Apart from the quality of the statistics (correlation coefficients in the appendix, Table A1), also the character of the relationship between the different component concentrations in the lacustrine sediment is affected by the prevailing synsedimentary environmental conditions. If, during the examination of the statistical relationships between two chemical substances, the component with the higher ionic potential is used as the dependent variable, the slope of the regression line steepens when the weathering intensity of the eroded matter goes up (Fig. 7). Regarding the statistical connection $Al_2O_3=f\{Fe_2O_3\}$, for stratigraphic unit 2, this is expressed in the steepening of the slope of the regression line by factor 1.15 compared with the counterpart relationship in unit 1 and by factor 1.58 compared with stratigraphic unit 3 (eq. 2-4). The same examination for the connection $Fe_2O_3=f\{K_2O\}$ establishes, for stratigraphic unit 2, a steepening of the slope of the

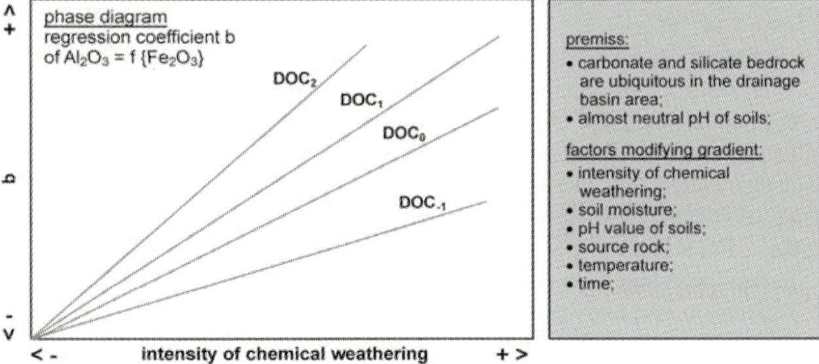

Fig. 7. Gradient of regression line of linear relation between Al_2O_3 and Fe_2O_3 contents of lacustrine sediments as paleoenvironment indicator

regression line by factor 1.28 compared with the corresponding relationship in unit 1 and by factor 1.43 compared with the same relationship in unit 3 (eq. 5-7). Proceeding from this comparison, allowing the same weathering and soil-formation processes the slope of the regression line is detrmined by the ionic potential of dependent and independent variables (ionic potentials: IP_{K+}=0.75, IP_{Fe3+}=4.7, IP_{Al3+}=5.9).

Because element mobility is due to the element‘s ionic potential, gaps between oxide concentrations in soils increase with weathering intensity (Krauskopf, 1967). While soils get eroded and form soil sediments, the steepness of the regression line’s slope between oxide concentrations in these deposits reflects the weathering intensity in the initial material.

5.2.4 Traces

The positively correlated connections between the heavy metals copper, chrome, nickel, and zinc and the silica, titanium, aluminium, and iron oxide contents of the lacustrine sediments (α<0.001) point towards the common origin of both substance groups from weathering and erosion of outcropping bedrock in the catchment of the Laguna de Jabonera. Due to the relatively high phyllosilicate contents in the lacustrine sediments of the Laguna de Jabonera, one can assume that the heavy metals are tied to the phyllosilicates – chiefly existing as illites – as exchangeable cations (Malle, 1990; Förstner et at., 1990; Schindler, 1991; Kühnel, 1992).

The positively correlated connections between the strontium and the sulphurtrioxide contents in the lacustrine sediments (α<0.001) in one respect follow from the common origin of these substances from the solution of the underlying Miocene evaporites in which strontium exists as strontium sulphate (Herrmann, 1961; for varying strontium cencentrations of stratighraphic units see chapter 5.1.3).

6 Conclusions

During the Holocene alternating subarid and subhumid climatic conditions resulted in the alternating predominance of various geomorphologic processes. In the Desierto de Calanda subarid climatic conditions led to extensive erosion processes and, thus, caused predominantly the deposition of detrital quartz, silicates, and carbonates in the basal part of the lacustrine sediments (stratigraphic unit 1). In opposite, overlying sediments (stratigraphic unit 2) are predominantly evaporitic, with quartz and silicates only as minor components - a sedimentary fabric due to a subhumid environment with processes of groundwater recharge and increased subsurface influx of aqueous solution. In the most recent sediments (stratigraphic unit 3) concentrations of silica and metal oxides again increase, and precipitation of sulphates is secondary. This change of decomposition is once more attributed to a change in the process system: As already pointed out, sedimentation of stratigraphic unit 2 was mainly influenced by subsurface inflow and reduced surface runoff, and so detrital deposits are only of secondary importance. In opposite, during deposition of the upper layer (stratigraphic unit 3) erosion dominated geomorphic process system occurred, possibly due to a sparse vegetation cover and to ephemeral to periodic heavy rains, corresponding to environmental conditions which altogether supported deposition of detrital sediments while the same time subsurface input of aqueous solution was reduced. At all, various chemical sediment characters (cf., strontium concentrations, quality of statistics) point out that environmental conditions during deposition of stratigraphic unit 1 were distinctly dryer than during deposition of stratigraphic unit 3. But, in general, comparison between stratigraphic units 1 and 3 is limited as for the most recent lacustrine sediments human impact has to be considered causing increased erosion rates (Stevenson et al., 1991). Increased detritals in the uppermost sediments are a result of these soil erosion processes (Faust and Diaz del Olmo, 1997).

Sediments in the depth of 50 to 60 cm, as well as in the depth of 260 to 270 cm indicate a short subhumid phase during the altogether subarid climatic condi tions throughout deposition of stratigraphic units 1 and 3. As pointed out by sediment chemistry, especially the deflection of the sulphur graph, these brief wetter phases resulted in a short-term increasing input of solutions by subsurface flow and, thus, in increased precipitation of evaporites.

For the lacustrine sediments of the Laguna de Jabonera investigated there are no data yet available. Comparison with investigations from neighbouring playa-type lakes, predominantly based on pollen analyses, allows to put Holocene environmental history of the Laguna de Jabonera into a preliminary time scale. The subhumid environmental conditions as reconstruted for the depositional phase of stratigraphic unit 2 correspond with the *late Subboreal*, while oldest lacustrine sediments are possibly from the *Atlantic-Subboreal* transition (Macklin et al., 1994; Davis, 1994). Based on this preliminary time-scale it also can be assumed that the short phase of increased humidity which caused deposition of high sulphate contents in 50 to 60 cm depth might be due to the *Modern Times* climatic pessimum of the *Little Ice Age*. A similar climatic trend in the most recent past was also found out by various authors investigating the sediments from the Laguna de Gallocanta, an endorheic basin c. 150 west of the Desierto de Calanda (Davis, 1994; Burjachs Casas et al.,1996; Schütt, 1998b).

Finally, it can be concluded that the chemical characteristics of lacustrine sediments yields valuable information about environmental conditions in the watershed. They indicate predominantly intensity of weathering and soil forming processes, but also give clues to intensity of erosion processes and input of solutes. These information supplement knowledge about the limnic environment during sedimentation which is derived from the mineralogical character of the sediment. The comparison with the results from other drainage basin areas shows that in order to interpret the chemical and mineralogical sediment features the individual petrographic, hydrologic, and geomorphic features of the drainage basin area must always be considered. By doing this, information about environmental change is gained which is as valuable as information derived from analysing pollen or microfossils.

Table 3. Laguna Jabonera/I: Summary of reconstruction of environmental conditions during deposition of lacustrine sediments (subdevided by core sections 1-3)

Limnic Environment		Drainage Basin Environment	Depth	Stratigr. Unit
ephemeral lake with hypersaline to saline environment	phase of increased inflow of groundwater and aqueous solutions 50 cm – 60 cm	subarid climate with low precipitation and high evaporation, low groundwater inflow and small input rates of aqueous solution, high rates of erosional processes cause predominance of detrital sediments, soil erosion processes	0 cm 165 cm	**3**
perennial or periodical lake with a saline to brackish environment		subhumid climate with relative high precipitation and reduced evaporation, groundwater recharge and high input rates of aqueous solution, relatively dense vegetation cover prevents erosional processes, intensified soil forming processes	166 cm 240 cm	**2**
ephemeral lake with hypersaline environment	phase of increased inflow of groundwater and aqueous solutions 260 cm – 270 cm	subarid climate with low precipitation and high evaporation, low groundwater inflow and small input rates of aqueous solution, high rates of erosional processes cause predominance of detrital sediments	241 cm 310 cm	**1**
		bedrock	311 cm	

Acknowledgements

From September 1994 until May 1996, the German Research Foundation (DFG) supported this project for obtaining proxy-data on Holocene climate change in northern and central Spain. William White helped to edit the language of this paper before submission.

References

Behbehani, A.-R., 1987. Sedimentations- und Klimageschichte des Spät- und Postglazials im Bereich der nördlichen Kalkalpen (Salzkammergutseen, Österreich). -*Göttinger Arbeiten zur Geologie und Paläontologie*, **34**.

Borch, C.C. van der, 1976. Stratigraphy and formation of Holocene dolomitic carbonate deposits of the Coorong area, South Australia. - *Jour. Sed. Petrology*, **46**, 952-966.

Borch, C.C. van der and Lock, D.E., 1979. Geological significance of Coorong dolomites. - *Sedimentology*, **26**, 813-824.

Burjachs Casas, F., Rodo, X. and Comin, F.A., 1996. Gallocanta: Ejemplo de secuencia palinológica en una laguna efímera. -Estudios Palinológicos (XI Simposio de Palinología, A.P.L.E.. 25-29.

Chesworth, W., 1992. Weathering systems. -In: I.P. Martini and W. Chesworth (ed.), Weathering, soils and palaeosoils. -Elsevier; Amsterdam, p. 19-40.

Davis, B.A.S., 1994. Paleolimnology and Holocene environmental change from endorheic lakes in the Ebro basin, north-east Spain. - Ph.D. Thesis, Newcastle Upon Tyne, U.K.

Degens, E.T., 1962. Geochemische Untersuchungen von Wässern aus der ägyptischen Sahara. -*Geol. Rundschau*, **52**, 625-639.

Dunne, T., Zhang, W. and Aubry, B., 1991. Effects of rainfall, vegetation and microtopography on infiltration and runoff. - *Water Resources Research*, **27**, 2271-2285.

Eckert, W., Nishiri, A. and Parparova, R., 1997. Factors regulating the flux of phosphate at the sediment - water interface of a subtropical calcareous lake: a simulation study with intact sediment cores. -*Water, Air and Soil Pollution*, **99**, 401-409.

Evans, R. and Kirkland, D.W., 1988. Evaporite environments as a source of petroleum. -In: Schreiber, B.C. (ed.), Evaporites and hydrocarbons. Columbia University Press; New York. p. 256-299.

Faust, D., 1995. Erkenntnisse zur holozänen Landschaftsentwicklung in der Campiñs Niederandalusiens. -*Geoökodynamik*, **16**, 153-171.

Faust, D. and Diaz del Olmo, F., 1997. Paläogeographie Südspaniens in den letzten 30000 Jahren: eine Zusammenstellung. -*Petermanns Geographische Miteilungen*, **141**, 279-285.

Flügel, E., 1978. Mikrofazielle Untersuchungen von Kalken. -Springer Verlag; Berlin.

Folk, R.L. and Land, L.S., 1975. Mg/Ca ratio and salinity: Two controls over crystallization of dolomite. - *Bull. Am. Assoc. Pet. Geol.*, **59**(1), 60-68.

Förstner, U., Ahlf, W., Calmano, W. and Kersten, M., 1990. Sediment criteria development. Contributions from environmental geochemistry to water quality management.

In: Heling, D. Rothe, P., Förstner, U. & Stoffers, P. (eds), Sediments and environmental geochemistry: selected aspects and case histories. Springer-Verlag; Berlin, Heidelberg, New York. p. 311-338.

Garcia de Pedraza, L. and Reija Garrido, A., 1994. Tiempo y Clima en España - Meteorologia de las Autonomas. - Madrid.

Håkanson, L. & Jansson, M., 1983. Principles of lake sedimentology. Springer; Berlin.

Heim, D., 1990. Tone und Tonminerale. Ferdinand Enke Verlag; Stuttgart.

Herrmann, A.G., 1961. Zur Geochemie des Strontiums in den salinen Zechsteinablagerungen der Staßfurt-Serie des Südharzbezirkes. -*Chemie der Erde*, **21**: 137-194.

Holser, W.T., 1979. Mineralogy of Evaporites. In: Burns, R.G. (ed.), Marine Minerals. - *Mineralogical Society of America, Short Course Notes* **6**, 124-150.

Horton, R.E., 1945. Erosional development of streams and their drainage basins: Hydrophysical approach to quantitative morphology. -*Geol. Soc. Am. Bull.*, **56**, 275-370.

Ibáñez, M.J., 1973. Contribución al estudio del endorreísmo de la depressión del Ebro: El foco endorreico al W. y SW. de Alcañiz (Teruel). -*Geographica*, **1**, 31-32.

Kelts, K. and Hsü, K.J., 1978. Freshwater carbonate sedimentation. -In: Lerman, A. (ed.), Lakes - chemistry, geology, physics. Springer Verlag; New York. p. 295-323.

Krauskopf, K.B., 1956. Dissolution and precipitation of silica at low temperatures. - *Geochim. Cosmochim. Acta*, **10**: 1-26.

Krauskopf, K.B., 1967. Introduction to Geochemistry. -McGraw-Hill Kogakusha Ltd.; Tokyo.

Kühnel, R.A., 1992. Clays and clay minerals in environmental research. -*Miner. Petrogr. Acta*, **25**, 1-11.

Langbein, R., Peter, H. and Schwahn, H.-J., 1981. Karbonat- und Sulfatgesteine. VEB Deutscher Verlag für Grundstoffindustrie; Leipzig.

Lerman, A., 1979. Geochemical processes - water and sediment environments. -John Wiley & Sons; New York. 481 p.

Macklin, M.G., Passmore, D.G., Stevenson, A.C., Davis, B.A. and Benavente, J.A., 1994. Responses of rivers and lakes to the Holocene environmental change in the Alcañiz region, Teruel, north-east Spain. In: Millington, A.C. and Pye, K. (eds), Environmental change in drylands: Biogeographical and geomorphological perspectives. John Wiley and Sons; London. P. 113-130.

Malle, K.-G., 1990. The pollution of the river Rhine with heavy metals. In: Heling, D. Rothe, P., Förstner, U. & Stoffers, P. (eds), Sediments and environmental geochemistry: selected aspects and case histories. Springer-Verlag; Berlin, Heidelberg, New York. p. 279-290.

Meyers, P.A. and Ishiwatari, R., 1993. The early diagenesis of organic matter in lacustrine sediments. -In: Engel, M.H. and Macko, S.A. (eds.) Organic geochemistry - Priciples and applications. -Plenum Press; New York, p. 185-209.

Morisawa, M.E., 1959. Relations of quantitative geomorphology to stream flow in representative watersheds of the Appalachian Plateau Province. -Tech. Rpt. No. **20**, Project NR 389-042, Office of Naval Research.

Morrow, D.W., 1979. The influence of the Mg/Ca ratio and salinity on dolomitization in evaporite basins. -*Bull. of Canadian Petroleum Geology*, **26**, 389-392.

Morrow, D.W., 1982. Diagenesis 2. Dolomite - Part 2: Dolomitization models and ancient dolostones. - *Geoscience Canada*, **9**, 95107.

Müller, G., 1997. Chronologie des anthropogenen Phosphor-Eintrags in den Bodensee und seine Auswirkung auf das Sedimentationsgeschehen. In: J. Matschullat, H.J. Tobschall and H.-J. Voigt (eds), Geochemie und Umwelt. Springer-Verlag; Berlin. p. 317-342.

Nathan, Y., 1984. The mineralogy and geochemistry of phosphorites. In: Nriagu, J.O. & Moore, P.B. (eds), Phosphate Minerals. Springer Verlag; New York. p. 275-291.

Riba, O., Reguant, S. and Vilena, J., 1983. Ensayo de síntesis estratigrafica y evolutiva de la cuenca terciaria del Ebro. In: Libro Jubilar J.M. Rios, II:131-159. IGME, Madrid.

Rogers, R.D. and Schumm, S.A., 1991. The effect of spoose vegetation cover on erosion and sediment yield. -*Journal of Hydrology*, **123**, 19-24.

Royse, C.E., Jr., Wadell, J.S. and Petersen, L.E., 1071. X-ray determination of calcite-dolomite: an evaluation. -*J. Sed. Petr.*, **41**, 483-488.

Sanchez Navarro, J.A, San Romàn, J. and Garrido, E., 1991. Las lagunas del sector Alcañiz-Calanda como una manifestación hidogeológica del drenaje de la cordillera ibérica en la depresión terciaria del Ebro. -*Boletin del Taller de Arquelogia Alcañiz*, **2**, 16-24.

Schindler, P.W., 1991. The regualtion of heavy metal concentration in natural aquatic systems. In: Vernet, J.-P. (ed.), Heavy metals in the environment. Elsevier; Amsterdam. p. 95-124.

Schröder, H.G., Windolph, H. and Schneider, J., 1983. Bilanzierung der biogenen Karbonatproduktion eines oligotrophen Sees (Attersee, Salzkammergut - Österreich). -*Arch. Hydrobiol.*, **97**, 356-372.

Schütt, B., 1998a. Chemical and mineralogical characters of lacustrine sediments as paleonvironmental indicators - An example from the Laguna Jabonera, Central Ebro Basin. -*Terra Nostra*, **98/6**, 115-120.

Schütt, B., 1998b. Reconstruction of Holocene Paleoenvironments in the Endorheic Basin of Laguna de Gallocanta, Central Spain by Investigation of Mineralogical and Geochemical Characters from Lacustrine Sediments. -*Journal of Paleolimnology*, **20**, 217-234.

Schütt, B., 1998c. Reconstruction of palaeoenvironmental conditions by investigation of Holocene playa-sediments in the Ebro Basin, Spain: Preliminary results. -*Geomorphology*, **23**, 273-283.

Sonnenfeld, P., 1984. Brines and evaporites. Academic Press; Orlando.

Stevenson, A.C., Macklin, M.G., Benavente, J.A., Navarro, C., Passmore, D. and Davis, B.A., 1991. Cambios ambientales durante el holoceno en el valle medio del Ebro: sus implicaciones arquelogicas. -*Cuaternario y Geomorfologia*, **5**, 149-164.

Stumm, W. and Leckie, J.O., 1971. Phosphate exchange with sediments: its role in the productivity of surface waters. -*Proceedings of the 5th International Water Pollution Research Conference*, 214-228.

Tennant, C.B. and Berger, R.W., 1957. X-ray determination of dolomite-calcite ratio of a carbonate rock. -*Amer. Min.*, **42**, 23-29.

UNEP, 1991. Status of desertification and implementation of the United Nations plan of action to combat desertification. -Report of the executive director to the governing council, 3rd special session. Nairobi.

Usdowski, E., 1973. Das geochemische Verhalten des Sr bei der Genese und Diagenese von Ca-Karbonat- und Ca-Sulfat-Mineralen. -*Contr. Miner. Petrol.*, **38**, 177-195.

Usdowski, E., 1974. Stabile und metastabile Reaktionen bei geochemischen Prozessen der Sedimentbildung und der Diagenese. - *Fortschritte der Mineralogie*, **52**, 81-93.

Vallentyne, J.R., 1962. Solubility and the decomposition of organic matter in nature. -*Arch. Hydrobiol.*, **58**, 423-434.

Appendix

Table A1. Laguna de Jabonera/I: Correlation matrix of major elements of core sections 1-3. Table shows adjusted r², asterix ** marks correlations below 95%-significance level

						section 1; [-310;-241] cm depth (n=7)					
		TiO_2	Al_2O_3	Fe_2O_3	MnO	K_2O	Na_2O	P_2O_5	CaO	MgO	SO_3
SiO_2	adj.r²=	0.69939	**	**	**	**	**	-0.51670	**	**	**
TiO_2	adj.r²=		**	**	**	**	**	**	**	**	**
$Al2O_3$	adj.r²=			0.6257	**	**	**	**	**	**	**
Fe_2O_3	adj.r²=				**	**	**	**	**	**	**
MnO	adj.r²=					**	**	**	0.62359	**	**
K_2O	adj.r²=						**	**	**	**	**
Na_2O	adj.r²=							**	**	**	**
P_2O_5	adj.r²=								**	**	**
CaO	adj.r²=									**	**
MgO	adj.r²=										**

						section 2; [-240;-166] cm depth (n=6)					
		TiO_2	Al_2O_3	Fe_2O_3	MnO	K_2O	Na_2O	P_2O_5	CaO	MgO	SO_3
SiO_2	adj.r²=	0.79543	0.72458	0.69742	**	0.59782	**	**	0.62055	**	0.59215
TiO_2	adj.r²=		0.98744	0.97230	**	0.90347	0.62085	**	0.72677	**	0.55795
$Al2O_3$	adj.r²=			0.99555	0.53347	0.95757	0.68336	**	0.76503	**	**
Fe_2O_3	adj.r²=				0.58106	0.97547	0.69841	**	0.80822	**	**
MnO	adj.r²=					0.56459	0.72521	**	0.59055	**	**
K_2O	adj.r²=						0.70129	**	0.83319	**	**
Na_2O	adj.r²=							0.51588	**	**	**
P_2O_5	adj.r²=								**	**	**
CaO	adj.r²=									**	**
MgO	adj.r²=										**

						section 3; [-165;0] cm depth (n=12)					
		TiO_2	Al_2O_3	Fe_2O_3	MnO	K_2O	Na_2O	P_2O_5	CaO	MgO	SO_3
SiO_2	adj.r²=	0.95638	0.84525	0.75639	**	**	**	**	**	**	0.91228
TiO_2	adj.r²=		0.80364	0.73942	**	**	**	**	**	**	0.84171
$Al2O_3$	adj.r²=			0.97843	0.46169	**	**	**	**	**	0.79997
Fe_2O_3	adj.r²=				0.55223	**	**	**	**	**	0.72444
MnO	adj.r²=					**	0.49389	**	**	**	**
K_2O	adj.r²=						**	**	**	**	**
Na_2O	adj.r²=							0.75358	**	**	**
P_2O_5	adj.r²=								**	**	**
CaO	adj.r²=									**	**
MgO	adj.r²=										**

Environmental changes in the Central Sahara during the Holocene — The Mid-Holocene transition from freshwater lake into sebkha in the Segedim depression, NE Niger

Roland Baumhauer[1], Erhard Schulz[2] & Simon Pomel[3]

[1] Department of Geography, Faculty of Geosciences, University of Würzburg, Am Hubland, D-97074 Würzburg, Germany

[2] Department of Geography, Faculty of Geosciences, University of Würzburg, Am Hubland, D-97074 Würzburg, Germany

[3] DYMSET/CNRS, University of Bordeaux III, F-33405 Talence Cedex, France

Abstract

The change from a fresh water lake to a sebkha during the middle Holocene was investigated in the Segedim depression/North-eastern Niger using continuous thin sections for micropetrography, palynology and for diatoms. This record is clearly divided into several units showing sequences of laminated anoxic to oxic clays, the stage of a sebkha with an inflow of loess, fine broken quartz grains and salts which are covered by dune sands made of rounded and clay covered quartz. The mineral assemblages of the three principal units are defined by the illite/kaolinite-calcite/aragonite-pyrite association (oxic lake), a kaolinite-calcite-halite-anhydrite (anoxic lake) and a montmorillonite-celestine-sodium carbonate-gypsum association connected to the sebkha. The composition of the sediments indicates a remarkable influence of sedimentation (ashes, phytoliths, charcoals, carbonates).

The diatom assemblages equally show the passage from fresh to saline water. The pollen spectra indicate the change from the Saharan savannah to desert vegetation.

These observations allow to distinguish the local and regional factors in the landscape ecology in order to detect the human impact and its influence on landscape changes in contrast to climatic changes. During the period of climatic degradation at about 6900 B.P., the influence of repeated fires could have had irreversible effects on the landscape formation.

Introduction

The regions near the Tropic of Cancer showed to be the most suitable areas to detect the evolution of landscape and climate. Recent investigations demonstrated the strong interaction on the monsoon and the Atlantic/Mediterranean front as well as the extension of the savannah systems up to 23°N in N-Mali (Petit-Maire, 1986; Schulz, 1994) during the middle Holocene. Surprisingly, the Sudanian savannahs were present at 19°N (Neumann, 1988) during that period and Sudanian alluvial vegetation types survived also during the Upper Pleistocene (Schulz et al., 1990), whereas at 25°N the vegetation showed Mediterranean influences during the Holocene (Wasylikowa, 1992).

In the heart of the Sahara, the depression of Segedim (NE Niger) shows a complicated landscape history during early and mid-Holocene. Depending on an isolated aquifer, a fresh water lake changed into a sebkha whereas under similar physical conditions in the depressions south of Segedim, the lake history continued up to the present time (Baumhauer & Schulz, 1984; Baumhauer, 1991).

In order to reconstruct the change from a fresh water lake to a sebkha environment more precisely, a core was analysed with help of continuous thin sections for the period of environmental changes, and a pollen and diatom analysis was carried out..

Geographical setting

The endorheic depression of Segedim lies at the northern margin of the Chad-Basin in North-eastern Niger (20°10′N 12°47′E) and is aligned between the Djado-Plateau and modern Lake Chad . It is typically located in the western foreland of a cuesta of karstified marine sand and siltstones of Senonian age (Faure, 1966). This forms a free-faced escarpment of an absolute height of 640 m a.s.l. and a relative height of 230 m. The Segedim depression consists of a Sebkha of approximately 10 km^2 with some groundwater inflow in the centre and at the eastern margin, stretching 2 km in an east-west direction with a maximum width of 7 km. Most of the depression is covered by a smooth sand layer. The surrounding plains are mostly covered with a serir formed by eroded gravels of some conglomeratic layers of the cuesta.

There is no climatic survey on the Segedim. In Bilma some 150 km to the south the climate is hyperarid and the annual precipitation does not exceed 10 mm with a potential evaporation of about 2700 mm/yr. These rare rainfalls are very irregular, mostly connected to the interaction of the monsoon and the polar front.

The present plant cover of the Kawar region shows the main characteristics of a desert vegetation (Boudouresque & Schulz, 1981). The permanent vegetation is contracted and restricted to wadis and depressions, whereas short-time grass and herb floras are connected to the aleatoric rainfalls. The depression of Segedim itself exposes a concentric mosaic of grass belts dependent on the structure of the sediments. An outer zone bears tussocks *of Panicum turgidum* on coarse sand and gravel, whereas the inner dune sand areas have stands of *Imperata cylindrica, Desmostachya bipinnata and Sprorobolus spicatus.* The marginal parts of the different subunits of the depression are colonised by trees *(Acacia raddiana, A. ehrenbergiana, Hyphaene thebaica). Phoenix dactylifera* grows on dune sands. The sebkha surface itself is plantless, but the salines and waterpoints are surrounded by *Tamarix canariensis and Juncus maritima.*

Coring

A first core (A) was already taken in 1981 using an open helix corer in the centre of the present sebkha. It indicated the passage of a fresh water lake to a sebkha environment during Mid- Holocene (Baumhauer & Schulz, 1984).

With the aim of a precise study of the transition zone, a second core (B) and a third core (C) were taken in 1989 and 1990 using a modified cullenberg corer at the same locality. Using a complete series of thin sections covering the transition zone, we established a microstratigraphical and micropedological non-destructive investigation on the changes during the passage from a lacustrine to a swamp environment, which changed to the sebkha and dune environment.

Dating

Two radiocarbon dates were obtained on the lower part of core A (Baumhauer & Schulz, 1984). The base of this core was dated on charcoals in lacustrine sediments at 790 cm depth to 7905 +/- 275 BP (Hv 114239), the transition to the sebkha environment at 590 cm depth was dated on charcoals too and showed an age of 6850 +/- 345 BP (Hv 11 424).

Micropetrographical characteristics

Unit I (719 to 680 cm)

This unit is composed of fine bluish to greyish clays intercalating regular ash layers. These clays are birefringent, punctuated or laminated and contain carbonaceous masses which are peptised by ashes. A rare mineral skeleton is made of fine quartz. Diatom frustules are frequent and algae cysts are present. This sequence is characterised by numerous charcoals and phytolithes, by melanised organic matter and pyrite. The punctured clay layers are carbonate rich. The carbonates are covered by clay minerals such as illite and kaolinite. The algo-bacterial laminae of the carbonates indicate fresh water conditions and oxic milieu. There is a certain rhythmic sedimentation of carbonated and non-carbonated clays.

Unit II (680 to 616 cm)

The sequence is composed of laminated/striated plastic ochre-reddish clays which are birefingent and pleochroic. The laminae also incorporate detritic quartz, diatoms as well as a planctonic gel. Charcoal particles are regularly present. The lacustrine milieu was very calm, indicated by the presence of face to face deposited kaolinites. The milieu became anoxydated and more and more saline, shown by the presence of halite, anhydrite replacing the pyrite as well as sodium carbonates replacing calcite and aragonites. The physical milieu is changing in the importance of the seasonal rhythmicity. It is shown by the transition from laminae to fine turbidites indicated by the masses of filtered clays. They derived from the erosion of soils rich in kaolinite caused by heavy rainfalls with large drops. There was a remarkable dust export from the clay horizons and the burned tropical fersialitic soils (cf. Felix-Henningsen, 1992)

Unit III (616 to 586 cm)

The sediment is composed of fine silts becoming coarser towards the top of the sequence. This loess is rich in detritic and pedological quartz and in salts (sulphates, gypsum, celestine and gay-lussite). These silts form a siltoskeletic plasma.

The evolution towards a sebkha is remarkable together with the aeolian input of loesses as well as from eroded upper soil horizons. This radical change is indicated by a reduction of the plant cover and a soil erosion accompanied by a change in the climatological conditions. The interaction monsoon/harmattan is enforced and the rain is characterised by big drops and tornadoes. This change in the ecological condition is manifested already early in the sequence.

Unit IV (586 to 561 cm)

This sequence is made up of a sapropel of finely broken quartzose sand. The sediment is rich in salt encrustation of gypsum, celestine and sodium carbonates. It also shows a strong vesicular porosity. The physical milieu became evaporative leading to a diminution of the water table and an evolution towards a sebkha. The input of quartz came from the corrasion of the bedrock representing an erosion of the soils to the R/C horizon. The climate degraded and it was characterised by a long dry season.

Unit V (561 to 542 cm)

The upper sequence is made up of a sapropel of coarse sand. The grains are covered by algo-bacterial films. Salts are present in the fine fractions as well as in the film cover. The physical milieu is marked by an important phase of dune mobilisation, deflation and the regression of the monsoon influence. Rainfall occurred only occasionally and the diminution of the vegetation cover is accompanied by a sand cover.

The upper part of' the record - not investigated with help of thin sections - is composed of several dune phases (Baumhauer and Schulz, 1984): Following unit V, coarse dune sands with brownish-reddish colour are present to 520 cm with some horizons of fine sand and brownish clays (520-475 cm) as well as brownish mid to fine sand with some coarser sands (475-420). Unit VI is made of yellowish mid to coarse dune sands (420-220) and mid to fine sands including lenses of coarse sand (220-90 cm). Unit VII is composed of yellowish coarse dune sand (90-60 cm) and coarse to fine sand (60-30 cm) topped by a salt crust including coarse sand (30-10 cm) and a surface crust 10-0 cm).

The nature of the laminae

The sediments of' core C and the lower part of core B consist of laminae and fine layered turbidites, whereas the upper part is dominated by turbidites of fine material washed in from the surrounding plains and in-blown dust. The lamination is not made by a sequence of organisms but by a fine grained matrix with quartz grains and diatoms instead intercalated by ashes and charcoals.

The laminae (838 cm) are composed of clayey and silty material intercalating regular ash layers. They include amorphous organic mater. Diatom frustules are frequent, but remarkably most of them are altered and sometimes degraded to amorphous siliceous material. Within these layers there is a characteristic presence of charcoal particles. Contrary to the punctual distributions of these elements, clays and silts show a characteristic criss-cross layering representing a calm and rarely interrupted sedimentation.

At about 683 cm this sedimentation type changes. Dark layers are represented now by pyrite in simple crystals or in framboid agglomerates. The package of clay and silt changes into a parallel face-to-face arrangement. Charcoal particles are rare but regularly present, and diatom frustules are more or less absent. The fine layering is made of fine grained turbidity, which reworked the fresh water sediments in the lower part over small distances before they showed their characteristic parallel arrangement in the clay and silt fraction.

In the upper sections the sediments are no longer laminated but still fine layered, representing the input of dust and fine grained turbidites. Loessic broken quartzes are intercalated with clay and silt material. Gypsum and halite crystals are present as well as pyrite.

The uppermost deposits represent the sebkha environment. They are composed of coated dune sand and a loessic matrix without any recognisable fine layering.

Diatom assemblages

The stratigraphical position of the samples analysed for diatoms, species having percentages of a minimum of >5 %, is given in Fig. 1. Percentage diatom analyses have been performed by counting 500 valves per sample. Diatom concentration varies from 0 to 106 valves x g^{-1}. Concerning the reconstruction of palaeoecological conditions from the diatom flora (salinity and chemical facies as well as life forms and habitat), the classification system according to Gasse (1987) and Gasse et al. (1987) is used in this paper. In the coarse detrital sequence of the upper part of the profile, diatoms have not been identified (layers 7, 8). Therefore these samples have been eliminated.

In the basal parts of the profile within the bluish clays of layers 14-16 (samples 27-29), the diatom association of *Fragilaria brevistriata F. construens, Melosira granulata and M. granulata var.* is abundant. Diversified freshwater to oligosaline epiphytic species as *Cyclotella ocellata, C. kuetzingiana, C. stelligera, Epithema zebra and Rhopalodia gibba* represent 1 to 5 % of the encountered forms. A rare species in Africa is *Navicula oblonga* (Schoemann & Meaton, 1982; Gasse, 1987), which is reported as an oligosaline form from lake Chad (Iltis, 1974; Gasse, 1987) and outside Africa it is recorded from alkaline, fresh to slightly Na-Cl-waters by Hustedt (1930, 1957) and Patrick & Reimer (1966).

The Melosira-Fragilaria ssp. (*Cyclotella ocellata)* association, in particular the rich frequencies of *Fragellaria brevistriata* (accompanied by *F. construens* in the upper part of the layers 14 and 15) are widespread in the diatom assemblages of Holocene lake deposits in the southern central Sahara, e.g. the former lakes in the endorheic depression in front of the cuestas (Servant-Vildary 1978; Baumhauer, 1986, 1990, 1991). *Fragilaria brevistriata* is regarded by Servant-Vildary (1978: 283) as a more oligotrophic form indicating well developed waterbodies while *F. construens* "est une *(F. brevistriata)* supplante dans les phases regressives"

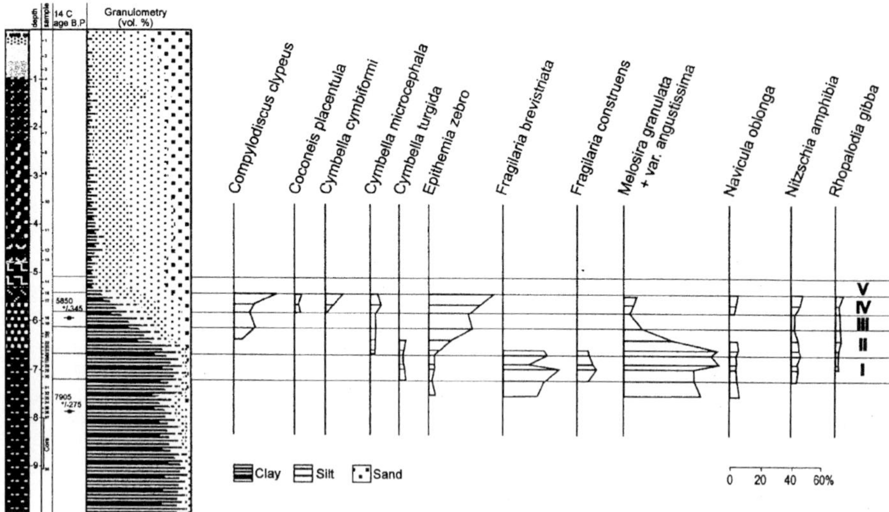

Fig. 1. Diagram of the Holocene diatom assemblages (Baumhauer 1991)

(Gasse, 1986, 1987) The afore mentioned author regarded comparable diatom assemblages as indicators of shallow dilute lakes, with slightly alkaline waters of the carbonate-bicarbonate type. *Cyclotella ocellata* and *C. kuetzingiana* may reflect an oligotrophic palaeoenvironment (Patrick, 1970) and the presence of *Rhopalodia gibba* is favoured by the abundance of macrophytes (Gasse, 1987).

In contrast to the previous stages, the diatom flora changes noticeably in the more silty clays on the layers 11-13 (samples 22-25) and the clayey-silty sands in layers 9 and 10 (samples 17-21). The relative abundance of *Melosira spp. (var. angustissima)* is associated with *Epithemia zebra,* according to Gasse (1987) an oligosaline periphytic-epiphytic species. *Epithemia* represents up to more than 40 % of the population in these layers and is accompanied by several oligo to mesohaline forms such as *Campylodiscus clypeus, Cymbella microcephala, Nitzschia amphibia* or *Rhopalodia gibba* as well as by 5-10 % *Melosira spp., Cymbella cymbiformis* and *C. turgida,* species most commonly found in freshwater. The relative abundance of *Campylodiscus clypeus* could be associated with a higher concentration and proportion of chloride and/or sulphate (Baudrimont, 1974). Gasse (1986, 1987) reported of shallow closed environments in the arid and semi-arid zones of Africa. The characteristics of the diatom associations in layers 9 and 10 may be attributed to a change from a shallow fluctuating palaeoenvironment to small pools, connected with seasonal fluctuations from fresh water to mesohaline conditions as well as episodic fresh water influx creating a stratified palaeoenvironment with superficial diluted waters.

Comparing the Holocene sediments, two successive waterbodies with different ecological conditions could be distinguished. In the Early Holocene, a freshwater lake of the (Ca + Mg) carbonate-bicarbonate-type with stable hydrological conditions, linked to a higher water table than today occupied the depression of Segedim. The chronological framework shows a relative rapid change in environmental conditions at about 6500 BP: the gradual desiccation from a shallow, freshwater lake over a short term period of small ephemeral ponds, containing a periphytic assemblage of saline, alkaline water, with a high content in chlorides or sulphates as well as carbonate content or superficial dilute water to a sebkha environment. Therefore in contrast to the southern part of the foreland depression of the cuesta of Bilma, the Kawar region, the Segedim Holocene lake dried up, whereas only 100 to 150 km away the lacustrine deposits indicate the persistence of previous (freshwater) conditions at least until 5000 BP (Servant, 1973/1983; Servant-Vildary 1978; Baumhauer, 1986, 1991).

The vegetation history

The palynological investigations of the different cores were made in order to detect the general landscape evolution of this central part of the Sahara and to answer the question whether there were any remarkable changes during the Holocene which could also be interpreted as climatically induced.

The pollen record is a composed record. It was established on cores B and C, for the gap between them we referred to the spectra of core A. This procedure was possible since the coring sites were close to each other and the upper parts of the two records were correlated by their pollen spectra (Schulz, 1994). The pollen spectra are predominated by the elements of an open vegetation. Trees, shrubs (including *Cheno-Am)* rarely reach 15 %. The pollen sum includes all elements with the exception of the spores. Any manipulation of the spectra was refused in order to show the local and regional vegetation and not to focus artificially on selected elements. The pollen diagram is simplified, rare elements are not mentioned in the diagram.

However, it is visible that the spectra are characterised by the continuous presence of *Acacia, Maerua, Capparis* together with *Combretum, Celtis* among the tree elements and *Cassia, Fagonia, Salvadora, Chorozophora, Guiera* as shrub elements. Together with the dominance of grasses and aquatics the herbs are represented by pioneer elements like Compositae, *Polycarpaea, Plantago or Mitracarpus.*

The diagram can be divided into five parts. It covers the period between the Early and Mid-Holocene.

A: Tree elements are dominated by *Acacia, Maerua,* together with a continuous presence of *Combretum, Celtis,* whereas the shrubs are characterised by *Fagonia, Cassia* also *Salvadora, Grewia and Guiera.* Elements like *Combretum* or *Guiera* may have been part of the long distance transport, but as close as 250 km to the South there was the combination of Saharan and Sudanian alluvial vegetation (Schulz et al., 1990). In comparison to the modern pollen spectra

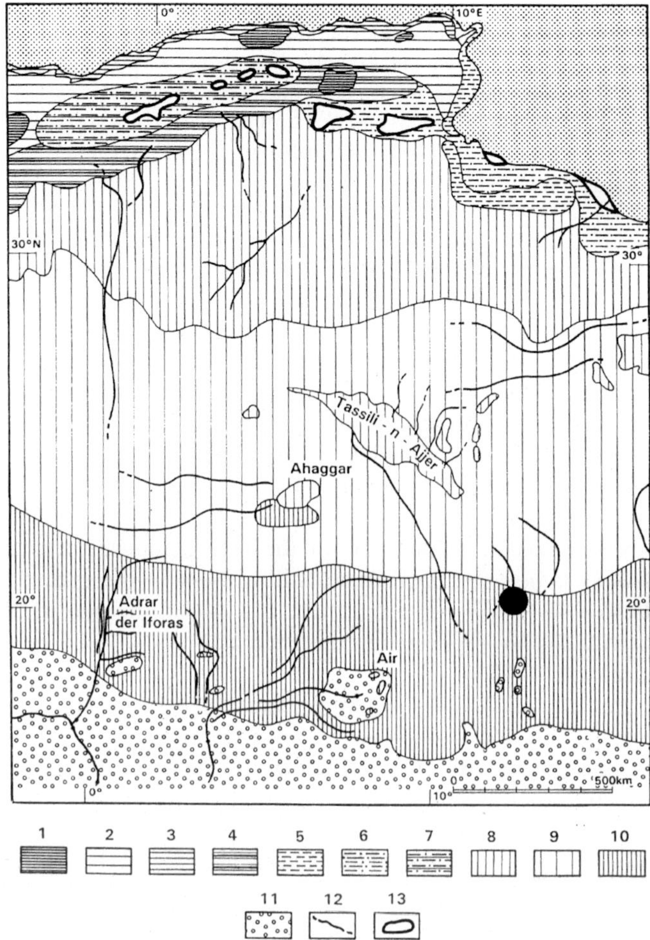

Fig. 2. Map of the Mid-Holocene vegetation of the Central Sahara and adjacent areas also showing the location of Segedim oasis (from Schulz & Adamou 1997):

<u>Meditarranaeis:</u>
1 coniferous forests (*Pinus, Cedrus*)
2 deciduous forests (*Quercus canariensis, Qu. pubescens*)
3 sclerophylluous forests (*Qu. canariensis, Qu. coccifera, Qu. suber*)
4 mixture of sclerophylluous and coniferous forests (*Qu. ilex, Pinus, Juniperus*)
5 sclerophylluous scrub (*Qu. coccifera, Olea-Pistacia, Erica, Arbutus*)
6 steppe (*Stipa, Lygeum*)
7 mixture of steppe and sclerophylluous forests (*Stipa, Lygneum, Qu. ilex, Pinus Juniperus*)

<u>Sahara:</u>
8 semidesert (*Artemisia Gymnocarpus, Ephedra, Chenopodiaceae, Stipagrostis*)

Fig. 2. cont.

	9	desert, predominant contracted vegetation (*Acacia-panicum, Tamarix-Stipagrostis*)
	10	saharan savannas (*Acacia-Maerua-Panicum*)
Sudan:	11	sudania savannas and tree formations (*Acacia, Combretum, Diospyros, Celtis*)
azonal, interzonal formations:		
	12	riparian vegetation (*Ulmus, Alnus, Salix / Ziziphus-Pistacia-Nerium*)
	13	scrub formation at chotts and coasts (*Chenopodiaceae, Limonium*)

(Schulz, 1990) it represents a sparse Saharan savannah vegetation perhaps also including some Sudanian alluvial elements like *Celtis* in the ravines of the cuesta.

B: This section shows a comparable mosaic of pollen spectra, but it is dominated by the Cyperaceae - both aquatic and dune plants. Thus it may result from a growth of the Cyperaceae belts of the lake.

C: Again, the spectra have a comparable composition like the ones in section A, but the Cheno-Am is rising to 20%.

D: The pollen spectra in this part of the record indicate a change in the vegetation. *Acacia, Maerua, Capparis* are still present in the spectra as well as *Fagonia, Salvadora or Securinega,* but there is a rise in the part of the aquatics this time also with *Typha*, representing the encroachment of reed belts into lake during its desiccation.

E: This part contains only a few and relatively poor spectra indicating the sebkha phase of the lake and an open vegetation Saharan type.

To sum up, one can state that the pollen spectra represent a similarity in their floristic composition, but they indicate a transition from the northernmost part of the Saharan savannah vegetation to the desert vegetation.The pollen spectra explain these changes from the diffuse to the contracted type of the *Acacia, Maerua, Cassia, Fagonia,* Gramineae vegetation and also the desiccation phase of the former lake into a sebkha environment which took place during the phases D and E.

Discussion on climatic change and the human impact

Local and regional factors

Local factors are strongly related to the depth of the water which reacts on the oxidising or reducing milieu and its extension via the riparian vegetation (Cyperaceae, Typha), which assured a filtration of the detritral inputs. Another factor of differentiation is the impact of fire on the landscape system. This could provoke an increase in grass cover and a diminution of the soil protection against erosion from rainfall and wind as well as cause local modifications in pH-value, water and sediment chemistry by the input of ashes.

A principal problem exists for the explanation of the origin - local or regional - of the clay sedimentation during unit I and II, or of the loessic phase of unit II as well as that of the sands during phase IV. A general problem is the reaction of the lake. Very early - compared to the other regions of the Bilma escarpment (Baumhauer, 1986) - the lake desiccated in the Segedim region. This could be a reaction to an isolated aquifer or to a more regional or global climatic change. During the lower part of this record, the pollen spectra as well as the mineralogical content reflect an interrelation between the (increasing) amount of elements transported over long distances and the expansive water surface.

Information on palaeoclimate

It is necessary to distinguish between the different stages of landscape development and those of the climate. In this case one can reconstruct two stages in the change of the lake situation during the transition of unit I to unit II. This represents the desiccation of the lake itself. The ecological situation in the Segedim depression is similar to that of the region .

From unit III, one has to count on an important change in the basin as well as in the region. Indicators of an efficient soil erosion, the input of freshly broken loess material together with the disappearance of open water imply a change in climatic conditions. One may think about a diminished vegetation and soil cover, which was no longer diffuse, and less important, but heavier rainfalls provoked strong erosion of the superficial soil horizons. Certainly, there was still that interaction of monsoon and Atlantic Mediterranean cyclons, but the rainfalls were frequently violent, indicating a stronger seasonality. During stage III an elevated evaporation created a salt accumulation and, consequently, the different aquifers (freshwater as well as salt water) individualised, caused by the diminution of the water surface, and they could create the different compartments of the depression. One also has to think about a relatively short-term and accentuated climatic change, because during period I the zone of the Sudanian savannahs did reach up to 19°N. The climatic gradient was more accentuated and caused another distribution of the landscapes.

The problem of fire

Charcoal particles and ashes are regularly distributed in the record. They are detected in the thin sections as well as in the pollen samples. The particles include a whole range between opaque black and sharp angular particles (cf. Clark et al 1989) as well as only partly burnt wooden material. We renounced a special counting of the charcoal particles in the various samples since their regular presence is the most important information. They are abundant during units I and III. A relative decrease in layers II and IV is attributed to the dilution by fine quartz which dominate the proportion of the charcoals and the ashes.

While it is clear that fire was a characteristic feature of the landscape's dynamic, the differentiation between accidental "natural" and man-made fires remains difficult. Accidental fires are well proved for the region to the South by fulgurites during the Mid-Holocene (Sponholz et al., 1993), but the regular presence of the charcoal and ash layers in the Segedim record make it difficult to explain it only by lightning. In contrast to this, one has to think on systematic fire setting as an intentional action of man. Fires caused by hunters may explain only a portion of these fires and regular occurrence in a milieu with an open tree cover is rare. One has to assume pasture or access fires to open water - certainly for small ruminants - since cattle keeping is only reported from about 4500 BP on in northern Niger (Gauthier, 1987).

Ashes and charcoal particles are also common for the lower fillings of the depressions of the cuesta plateau in Segedim as well as in the Kawar (Sponholz, 1991). Sponholz points out that these charcoal particles are not washed in from younger sediments above and that they are younger that 4000 BP.

These fires could well destroy a certain part of the plant and soil cover especially in the Cyperaceae- and Typha -belt around the lake and they could provoke an important mobilisation of fine-grained material of the superficial horizons (clay within units I to II and silts in unit III). A similar effect of human impact on vegetation, soils and landscape system has also been observed in northern Sudan (Haynes et al., 1989).

The presence of neolithic and epipalaeolithic tools in that region confirm a human occupation (Tillet, 1983; Baumhauer, Morel & Tillet, 1991). Observations on the human impact on the southern boundary of the Sahara (Smith, 1980; Wilson, 1982; Haynes et al., 1989; Klute, 1992) indicate an instrumentalisation of fire and also a first stage in the domestication of the landscape (Yen, 1989). These human impacts would certainly have had more important effects during the periods of degradation of the climate, even if these actions were not very intensive. Perhaps they were localised and/or accentuated in the forelands of the cuestas.

The observations concerning the role of fire in the semihumid and semiarid tropical regions (Goldammer, 1990) show several effects: a certain monotonisation of the species diversity by the support of the Gramineae and the pyrophytes. There is a diminution of organic matter in the soils and a mineralisation of biomass (as is shown in the Segedim core) as well as an augmentation of the mobility of surface horizons caused by the denudation and augmentation of fine grained particles as an effect of ash dispersion.

Conclusion

The different investigations on the Segedim record revealed three main results.

A. The transition of an Early to Mid-Holocene groundwater-fed freshwater lake to a swamp environment and consecutively to a sebkha suggests a more complex influence of local (and regional) factors (e.g. geomorphology, hydrology, hydrogeology) than a mere climatic dependence. The stratigraphical, sedimentological and ecological status obtained for several endorheic depressions in the region seems to be mainly a reflection of differences in groundwater catchment, aquifers of different size, local topographical conditions and changes in geomorphology superimposed on the major climatic tendencies (cf. Baumhauer 1991).Even a gradual degradation of the climate and a final shift to the predominance of the monsoonal rainfall regime took place around 6500 BP - as demonstrated by the change of the Saharan savannah into desert vegetation – cannot explain the differences in behaviour of the individual palaeolakes in the region: waterbodies (e.g. Segedim) with clear changes in water balance and water chemistry occurred, while perennial freshwater lakes (e.g. Kawar) persisted nearby (cf. Baumhauer 1986,1991, Baumhauer et. al. 1991/1997).

B. The laminae of the lower part of the record consist of fine grained filtered clay and silt material representing regular dust input. Charcoal particles and diatom frustules in different stages of alteration are regularly distributed in this matrix, ash and pyrite layers intercalate the clay and silt sedimentation. These laminae may be seasonal but not necessarily annual. They are also intercalated by fine layered turbidites. The change to swamp and sebkha environment is proved by the predominance of fine grained turbidites and the growing mass of loess and dune sand accumulation together with a rising salt content.

C. The regular presence of charcoal particles and ashes point to an early stage of human impact on the landscape's dynamic as well as on the development of the lake's nature itself.

References

Baudrimont, R., 1974, Recherches sur les diatomées des eaux continentales de l'Algerie : écologie et paléoécologie. Mém. Soc. Hist. Nat. Afrique du Nord, 12, N.S. 249 p.

Baumhauer, R., 1986, Zur jungquartären Seenbildung im Bereich der Stufe von Bilma (NE-Niger). Würzb. Geogr. Arbeiten., 65, 235 p.

Baumhauer, R., 1990, Zur holozänen Klima-und Landschaftsentwicklung in der zentralen Sahara am Beispiel von Fachi/Dogonboulo (NE-Niger). Berliner Geographische Studien, 30, pp. 35-48.

Baumhauer, R., 1991, Palaeolakes of the south central Sahara - problems of the palaeoclimatological interpretation. Hydrobiologia, 214, pp. 347-357.

Baumhauer, R., Morel, A. & Tillet, Th. 1991/1997, Air-Ténéré-Djado-Kawar: In Sahara. Paléomilieux et peuplement préhistorique au Pléistocéne supérieur. Colloque de Solignac, Univ. Limoges, ch. 4 B, 35 p; Palaeoenvironments and Prehistoric Populations in the upper Pleistocene, L'Harmattan, Paris, 229-266.

Baumhauer, R. & Schulz, E., 1984, The Holocene lake of Segedim, Kaouar, NE - Niger. Palaeocol. of Africa, 16, 283-290.

Boudouresque, E. & Schulz, E., 1981, The flora and vegetation of NE-Niger (Djado, Kaouar and Ténéré). Willdenovia, 11, pp. 363-394.

Clark, J. S., Merkt, J. & Müller, H., 1989, Post-glacial fire, vegetation and human history on the northem alpine forelands, South-West Germany. J. Ecol. 77, 897-925.

Faure, H., 1966, Reconnaissance géologique des formations sédimentaires post-paléozoiques du Niger oriental. Dir. Mines et Géol. Bureau de Rech. Géol et Min., Paris, 1, 630 p.

Felix-Henningsen, P., 1992, Merkmale, Verbreitung und klimazonale Ausprägung frühholozäner Feuchtzeitböden in der Ténéré, Ost-Niger. Würzb. Geogr. Arbeiten, 84, 97-129.

Gasse, F., 1986, East African diatoms. Taxonomy, ecological distribution. Bibliotheca Diatomologica, 11, 202 p.

Gasse, F., 1987, Diatoms for reconstructing palaeoenvironments and palaeohydrology in tropical semi-arid zones. Hydrobiologica, 154, 127- 163.

Gasse, F., 1988, Diatoms, palaeoenvironments and palaeohydrology in the western Sahara and the Sahel. Würzb. Geogr. Arb., 69, 233-254.

Gasse, F., Fontes, J.Ch., Plaziat, J.C., Carbonnel, P., Kaczmarzka,I., De Dekker, P., Soulie-Märsche, I., Callot, Y. & Dupeuple, P.A. 1987, Biological remains, geochemistry and stable isotopes for the reconstruction of environmental and hydrological changes in the Holocene lakes from North Sahara. Palaeogeography, Palaeoclimatology, Palaeoecology, 60,1-46.

Gauthier, A., 1987, Prehistoric Men and Cattle in North Africa: A Dearth of Data and a Surfeit of Models. In: Close, A.E. (ed.), Arid North Africa, Essays in Honour of Fred Wendorf. S.M. U. Press, Dallas, 163-187.

Goldammer, J.G. (ed.), 1990, Fire in the Tropical Biota. Ecosystem Processes and Global Challenges. Springer, Berlin, 497 p.

Haynes, C. V., Eyles, C. H, Pavlish, L. A., Ritchie, J. C. & Rybak, M., 1989, Holocene palaeoecology of the eastern Sahara: Selima oasis. Quaternary Science Reviews, 8, 109-136.

Hustedt, F., 1930, Baccilariophytha (Diatomae). In Pascher, A. (ed.), Die Süßwasserflora Mitteleuropas, 10, Fischer, Jena 466 p.

Hustedt, F., 1957, Die Diatomeenflora des Flußsystems der Weser im Gebiet der Hansestadt Bremen. Abh. Naturwiss. Ver. Bremen, 34, 18-140.

Iltis, A., 1974, Le phytoplancton des eaux natronées du Kanem (Tchad). Influence de la teneur en sels dissous sur le peuplement. Thèse, Univ. Paris VI, 313 p.

Klute, G, 1992, Die schwerste Arbeit der Welt. Alltag der Tuareg-Nomaden. Trixter, München, 250 p.

Neumann, K., 1988, Die Bedeutung der Holzkohlenuntersuchungen für die Vegetationsgeschichte der Sahara - das Beispiel Fachi - Niger: Würzburger Geogr. Arb, 69, 71-85.

Patterson,W.A., Edwards, K.J. & Maguire D.J, 1987, Microscopic charcoal as a fossil indicator of fire. Quaternary Science Revies, 6, 3-23.

Patrick, R., 1970, The diatoms. Trans. A, Phil. Soc., 60, 112-120.

Patrick. R. & Reimer, C.W., 1966, The diatoms of the United States. Vol I Monogr. Acad. Nat. Sci. Philadelphia, 13, 688 p.

Petit-Maire, N., 1986, Paleoclimate in the Sahara of Mali: a multidisciplinary study. Episodes, 9/1, 7-16.

Pomel, S. & Otto, Th., 1994, Origine et fonctionnement des sols de la région de Maroua-Salak (Nord Cameroun). In Press.

Pomel, S. & Schulz, E. 1992, Les sols des savannes anthropogènes du Cameroun. Würzb. Geogr. Arb., 84, 263-288.

Schoeman, F. R. & Meaton, V.H., 1982, Catalogue of recently described diatoms taxa from Africa and neighbouring islands (1965-1980). CSIR Special report WAT 64, Pretoria 64 p..

Schulz, E., 1990, Zwischen Syrte und Tschad. Der aktuelle Pollenniederschlag in der Sahara. Berliner Geographisches Studien, 30, 193-220.

Schulz, E., 1994, The southern limit of the Mediterranean vegetation in the Sahara during the Holocene. Hist. Biol. 9, 137-156

Schulz, E. & Adamou, A., 1997, Die Grenzen der "neolithischen" Revolution. Gab es einen frühen Ackerbau in der Sahara? Würzb. Geogr. Arb.. 92, 71-96

Schulz, E., Joseph, A. Baumhauer, R., Schultze, E. & Sponholz, B., 1990, Upper Pleistocene and Holocene history of the Bilma region (Kawar, NE - Niger). In: Recent data in African Earth Science. Proc. 15. Coll. Afr. Geology, Nancy, Publ. CIFEG, Paris, 22, 281-284.

Schulz, E. & Merkt, J., 1996, Transsahara. Die Überwindung der Wüste. Würzb. Geogr. Manuskripte, 38, 116 p.

Servant, M., 1973/1983, Séquences continentales et variations climatiques: évolution du bassin du Tchad au Cénozoique supérieur. Thèse Univ. de Paris, 348 p. Trav. Doc. ORSTOM, 159,573 p.

Servant-Vildary, M., 1978, Etude de diatomées et paléolimnologie du bassin tchadien au Cénozoique supérieur. Trav. Doc. ORSTOM. 84, 346 p..

Smith, R., 1980, The environmental adaptation of nomads in the West African Sahel. A key to understanding prehistoric pastoralists. In: Williams M.J. & H. Faure (eds): The Sahara and the Nile. Rotterdam, pp. 476-488.

Sponholz, B. 1991: Sedimentologische Untersuchungen an Verfüllungen von abflusslosen Hohlformen im Nordost-Niger. Bericht DFG, Würzburg, 21 p.

Sponholz, B., Baumhauer, R., & Felix-Henningsen, P., 1993, Fulgurites in the south central Sahara, Republic of Niger, and their palaeoenvironmental significance. The Holocene, 3,2,97-104.

Tillet, Th., 1983, Le Paléolithique du Bassin tchadien -septentrional (Niger-Tchad). Ed. CARS, Paris, 319 p..

Wilson, R.T., 1982, The "gizu" : Winter grazing in the south Libyan Desert. Journal of Arid Environment, 1,325-342.

Wasylikowa, K., 1992, Holocene flora of the Tadrart Acacus area, SW Libya, based on plant macrofossils from Uan Muhuggiag and Ti-n-Torha/Two caves archaeological sites. Origini, XVI, 1233-152.

Yen, D.E., 1989, The domestication of environment. In: Harris, D.R. & Hillman, C.C. (eds.) Foraging and Farming. Unwin Hyman, London, 55-78 .

Genesis and Paleo-ecological Interpretation of Swamp Ore Deposits at Sahara Paleo-lakes of East Niger

Peter Felix-Henningsen

Institut für Bodenkunde und Bodenerhaltung
Heinrich-Buff-Ring 26, D-35392 Gießen
Germany

Summary

In formerly vegetated flat lake-shore areas of Pleistocene and Holocene paleo-lake depressions in the Sahara of East Niger (Ténéré, Tchigai mountains and in the Erg of Bilma), ancient dune sands are covered by rampart-like or flat beds of individual or networked rhizoconcretions. The massive goethite accumulation, which partly includes an outer fringe of lepidocrocite, impregnated the ancient dune sands. Apart from Fe, P, Ca, and Mg, other heavy metals were also concentrated. The formation and morphological differentiation of these "swamp ores" were generally bound at vegetated shallow water areas of paleo-lakes in ancient dune fields. Accordingly, the swamp ores of the Ténéré, which has flat to undulating relief, display a large dissemination. In contrast, in the Erg of Bilma the high altitude and steep slopes of ancient dune ridges led to steeper shore areas of the paleo-lakes, at which beds of rhizoconcretions were unable to develop.

The oxides were formed by oxidation of Fe^{2+}-ions from the lake water and concentrated around the roots in the upper root zone of the swamp vegetation. The lack of oxygen in the warm lake water of the shore region, as well as the decomposition of vegetation residues, excluded high redox potentials within the deeper water near the subhydric soil surface. Hence, the formation of rhizoconcretions can only be explained by the specific physiological characteristics of the swamp vegetation, which was able to supply oxygen to the roots through an aerenchyma. The release of surplus oxygen from such roots obviously caused high redox potentials at the root surface and in the neighbouring root environment. As a result precipitation of Fe and Mn oxides occurred, which adsorbed nutrients and heavy metals from the soil solution. The redistribution of the ions from the reduced sediments of the lake basin into the root zone of the shore area resulted from diffusion and mass flow. Paleo-climatically, the swamp ore deposits denote humid periods accompanied by a stable lake water-table over a long period of time. The sequence of several swamp ore beds along the former shore, at different elevations above the lake floor, is evidence of decreasing paleo-lake water levels within a period of increasing aridity.

Zusammenfassung

"Sumpferze" im Uferbereich von Paläoseen der Sahara Ost-Nigers

In den ehemaligen Flachuferbereichen von pleistozänen und holozänen Paläoseen der Ténéré, des Tchigai-Berglandes und im Erg von Bilma sind im obersten Horizont von vergleyten Altdünensanden wallartige und flache Lagern aus einzelnen oder miteinander vernetzten Rhizokonkretionen ausgebildet. Sie bestehen aus massiven Goethitanreicherungen, z.T. mit einem äußeren Saum aus Lepidokrokit, die den Altdünensand imprägniert haben. Neben Fe wurden P, Ca, Mg und Schwermetalle angereichert. Die Entstehung und morphologische Differenzierung dieser "Sumpferze" war regelhaft an vegetationsbesiedelte Flachwasserbereiche (Untiefen und Flachuferzonen) der Paläoseen in den Altdünengebieten gebunden. Ihre morphologische Differenzierung in massive Lager aus verbackenen Rhizokonkretionen einerseits und einzeln stehende Oxidstengel erfolgte mit zunehmender Wassertiefe. Daher weisen Sumpferze an Paläoseen in der flachwelligen Ténéré eine große Verbreitung auf, in dem durch steilere Altdünenzüge und Uferzonen geprägten Erg von Bilma sind sie dagegen relativ kleinräumig verbreitet. Die Oxide entstanden durch Oxidation von Fe^{2+}-Ionen des Seewassers und der Oxidkonzentration um die Wurzeln im oberen Wurzelraum der ufernahen Vegetation. Da die geringen Sauerstoffgehalte in dem warmen Seewasser der Uferregion sowie die Sauerstoffzehrung durch den Abbau von Vegetationsrückständen hohe Redoxpotentiale im bodennahen Seewasser ausgeschlossen haben, kamen nur die spezifischen physiologischen Eigenschaften der Sumpfvegetation, die über ein Aerenchym zur Versorgung der Wurzeln mit Sauerstoff verfügte, für die Bildung der Rhizokonkretionen in Frage. Die Freisetzung von überschüssigem Sauerstoff aus den Wurzeln hatte hohe Redoxpotentiale und damit eine Oxidabscheidung in der unmittelbaren Wurzelumgebung zur Folge. Die Umverteilung der Ionen aus den reduzierten Sedimenten des Seebeckens in den Wurzelraum der Uferzonen erfolgte über Diffusion und Massenfluss.

Paläoklimatisch kennzeichnen die Sumpferzlager jeweils lange Perioden mit gleichmäßigen Seespiegelständen. Ihr Vorkommen in unterschiedlichen Höhenlagen über dem Seeboden an flachen Uferzonen belegt den Trend sinkender Wasserstände der Paläoseen infolge der zunehmenden Aridität des Klimas und unterbrochen von humideren Intervallen.

1 Introduction

In extremely arid to semiarid climatic regions of the Sahara desert, fossil and relict paleosols serve as evidence of past humid climatic periods. These soils developed in eolian sands and can be used to indicate the age and distribution of ancient dunes, formed during dry climatic phases which occurred prior to development of the paleosols in the humid phases, as well as the extension of a humid climate during paleo-monsoon events in what are now desert centres. From previous investigations, three humid periods during the upper Quaternary must be considered in the southern Sahara of East Niger:

a) an Upper Pleistocene (Ghazalien) humid period between 40,000 and 20,000 BP (Servant, 1983),
b) an late Pleistocene to early Holocene (Tchadien) humid period between 14,000 - 7,500 BP with a maximum of humidity between 10,000 and 7,500 BP (Servant 1983), and
c) a middle Holocene (Nouakchottien) humid period (Young Neolithic) between 4,500 - 3,000 BP (Michel 1973).

In the Sahara of East Niger, mainly in the Ténéré and the Grand Erg de Bilma, relict and fossil soils on ancient dunes are associated with lacustrine sediments of paleo-lakes, which during the early and middle Holocene extended in spacious shallow depressions (c.f. Baumhauer 1991, 1993; Felix-Henningsen 1992, 2000; Grunert 1988; Völkel 1988, 1989).

Paleosols were investigated at selected sites along a 600 km long SW-NE transect, extending from the Air Mountains across the Ténéré desert to the Tchigai Highlands (Fig. 1). Paleopedological work in this region was last carried out by Skowronek (1988) near Seguedine on the northern Kaoar escarpment. The investigations along the transect across the Ténéré mainly focused on paleosols developed on ancient dunes of Pleistocene age, which are extensive in the study area. Catenas of late Pleistocene to early Holocene fossil soils on ancient dunes, covered by a centimetre to decimetre thick layer of modern eolian sand, systematically display a sequence of (Chromi-) Cambic Arenosols, developed on plateaus and slopes of ancient dune ridges, Gleyic Arenosols near the paleo-lake shores, and white, bleached dune sands of submerged soils of the paleo-lake bottom covered by layers of lacustrine sediments from carbonate, diatomite and silt, often interspersed with organic matter (Felix-Henningsen 2000). A major part of the la

Fig. 1. Location map showing study sites under investigation in the Sahara desert of East Niger. Only the numbered sites for Ténéré (6 and 8) and the Tchigai mountains (11) are of interest in this contribution

custrine sediments has been eroded by deflation in subsequent arid climatic phases.

The massive beds of concretionary iron oxides, which were protected from deflation, occur as dam-like ridges or concentrations of autochthonous, vertically standing cauliform concretions, as well as a strew of concretions on the surface of the ancient dunes systematically following the course of flat lake shores. As they always form a part of the surface horizon of a Gleyic Arenosol with strong bleaching of the lower horizons and weak iron oxide mottling of the uppermost horizon, they can be interpreted as a part of the near-shore submerged paleosols. Such banks of massive and hard oxide concentrations referred to in this paper are called "swamp iron ores". Terms for single cauliform oxide concentrations, which occur in hydromorphic soil horizons, were named rhizo-concretions or root tubules (Schwertmann & Taylor 1989) and pipestems (e.g. Fitzpatrick 1988, Schwertmann & Fitzpatrick 1992). As they always occur on top of ancient dunes of Pleistocene age, the swamp iron ores cannot be confused with pre-Quaternary lateritic crusts that frequently occur on top of hard sedimentary rocks of Cretaceous age as described by Schwarz (1992). Previous descriptions of such Saharan swamp iron ores are meagre, and specific investigations and paleo-ecological interpretations in the context of paleopedology are missing. Only Maley (1981), Williams et al. (1987) and Baumhauer (1993) describe the occurrence of beds of swamp iron ores in the Sahara of East Niger, while Kröpelin (1993) investigated similar beds of rhizo-concretions along shores of paleo-lakes of the northwestern Sudan.

This paper discusses the characteristics, genesis and paleo-climatic implication of swamp iron ores of the Ténéré and southern Tchigai Highlands (Fig 1). These iron ores are a part of paleosols that, in all probability, developed during the Early Holocene humid period (c.f. Felix-Henningsen 2000). Because absolute dating of the paleosols was not possible in this study, their relative ages were estimated by comparing soil type and degrees of weathering. Interpretations incorporated results from geomorphological and archaeological investigations, which were carried out at the same locations, and from previous paleopedological and geomorphological investigations in the Grand Erg de Bilma, adjacent to the Ténéré, and the Sahelian zone of East Niger, carried out by Grunert (1988), Völkel (1988, 1989) and Völkel and Grunert (1990). Radiocarbon dating of organic and calcareous deposits adjacent to swamp iron ores of Ténéré paleo-lakes establish the beginning of the lacustrine and diatomitic sedimentation to be during the early Holocene (8.7 – 9.0 ka BP) and a continuation until the lakes changed to swamps at 5.5 – 6.0 ka BP (Baumhauer 1993).

2 Study area and methods of investigation

2.1 Study area

The geological and geomorphological development of East Niger, as well as the Quaternary climatic history, were comprehensively summarised by Grunert (1988), Baumhauer et al. (1989), Sponholz (1989), Völkel (1989) and Pfeiffer (1991). In the Ténéré, Upper Cretaceous sandstones, with intercalated glauconitic layers of the basin of Bilma, form small mesas, plateaus and escarpments, which are orientated along N - S trending fault zones. The bedrock is covered by Pleistocene eolian sands of ancient ergs and recent dune fields, fluvial sediments of the wadis, clayey to silty middle terrace sediments of the Air piedmont plain, and lacustrine sediments of paleo-lakes (Baumhauer, 1986, 1991, 1993; Baumhauer et al., 1989, Pfeiffer and Grunert, 1989; Skowronek, 1979). The Pleistocene dunes were stabilised by the development of paleosols and display a flattened relief because of denudation that occurred at the beginning of each humid period. Where the paleosols were truncated in subsequent arid periods, the Pleistocene dune sands still display an increased bulk density and slight to moderate induration caused by accumulation of soluble salts, carbonate and amorphous silica, which were leached downwards from the paleosols. On the other hand, the modern dunes consist of highly mobile uncemented sand, accumulated in flat sand sheets with ripple marks and high dune ridges with pronounced crests following the direction of the trade winds.

The Air Mountains, with elevations of up to 2,000 m a.s.l., and the Tibesti (Chad), with elevations of up to 3,300 m a.s.l., consist of Precambrian metamorphic and volcanic rocks. The plains of the Ténéré have an elevation of approximately 400 m a.s.l., and the sandstone plateaus of the southern Tchigai Highland rise to 500 m a.s.l.. South of the Agadez-Bilma line, the Ténéré passes into extensive fields of active longitudinal dunes known as the Grand Erg de Bilma. Climatically the area is classified as extreme desert. Episodic monsoonal summer precipitation leads to long-term average values of about 145 mm yr^{-1} at Agadez, 20 mm yr^{-1} at Bilma, and 10 mm yr^{-1} at Djado in the north of the area studied (Sponholz, 1989).

2.2 Methods of investigation

At selected sites (Fig. 1) along the transect, paleosols at or near the present land surface were exposed by digging. Soil horizons were described (Munsell colour, texture, structure and bioturbation, hardening and cementation, calcification and gleyification) and sampled for further pedochemical, physical, mineralogical and micromorphological investigations.

All chemical and physical soil characteristics were analysed according to methods described by Schlichting et al. (1995). Main and trace elements of the swamp iron ores and rhizo-concretions were investigated on ground samples by X-ray fluorescence.

3 Regional occurrence and characteristics of swamp iron ores

3.1 General characteristics

During the early Holocene humid period, extended freshwater-lakes existed throughout the Sahara of East-Niger, some with diameters of many kilometres and depths from several meters to almost 40 m (see Baumhauer 1991, 1992; Grunert et al. 1991). Extended basins of paleo-lakes within the even to flat-wavy landscape of the Ténéré are characterised by lacustrine sediments and white, bleached ancient dune sands. On top, and partly within the uppermost horizon of Gleyic Arenosols from ancient dune sand in the former flat-shore-areas, beds of stem-like, branched, networked and cemented rhizo-concretions from iron oxide form shallow ridges or flat deposits on top of the ancient dunes. Autochthonous beds of swamp iron ores are up to 40 cm thick, while the diameter of a single cauliform rhizo-concretion varies between 1 - 5 cm. The bottom of such massive banks of cauliform rhizo-concretions is embedded within the weathered and mottled Bwg horizon of the ancient dune, which is only weakly cemented by an accumulation of amorphous silicic acid. The rhizo-concretions consist of massive, concretionary accumulation of goethite without Al substitution, which impregnated the ancient dune sand, concentric around a central root- or stalk-canal. The dun coloring of the oxides around the middle canal indicates a stronger enrichment of Mn oxides. The hard rhizo-concretion proceeds into the adjacent, unconsolidated dune sand with a soft transition zone of a few mm thickness and with a decreasing accumulation of oxides. In this zone, orange coloured oxides show, according to X-ray analyses, a higher proportion of poorly crystallized lepidocrocite.

Photo 1. Ténéré, site 6 (c.f. Fig. 1), ancient dune with an autochthonous fossil, reddish brown Chromi-cambic Arenosol of the early Holocene humid period, dissected by animal burrows and covered by modern eolian sand

Photo 2. Ténéré, site 8 (c.f. Fig. 1), ancient dune with an autochthonous fossil, bleached Gleyic Arenosol developed at the bottom of a paleo-lake of the early Holocene humid period, covered by a thin layer of silty lacustrine sediments and modern eolian sand. The thick Ah horizon and the crotowina possibly developed after the end of the lake period, when the environment was still rather moist

Photo 3. Massive swamp ore deposit formed by cementation of single rhizo-concretions near the surface of a fossil Gleyic Arenosol, in vertical direction about 30 cm thick, surrounded by fragments of swamp ore (Ténéré, site 6, c. f. Fig. 1)

The swamp-ores were primarily exposed by deflation. Beds with a strong oxide-enrichment and crust-like cementation thus form 1 - 3 m wide ramparts, strictly following the former shore lines of the paleo-lakes. They occur mainly in bays with a flat shore and inclinations < 10°. They are several meters to decametres wide and extend from some meters to several hundred meters along the shore line. At more inclined paleo-lake shores, the swamp-ore beds become increasingly narrow or are not developed.

The massive cementation of the oxide stalks ends abruptly towards the lake basin. They occur as an edge of single-oxide stalks, with increasingly further distance becoming thinner and shorter, and finally disappearing after some meters to decametres (Photo 3). Through deflation, the oxide-stalks were exposed and fractured by corrasion such that the paleo-lake bottom near the former shore line is covered by cm- to mm-large stalk fragments.

In the direction of the paleo-lake basin the beds of swamp iron ore end with a clear border and are replaced by a zone of soft lake marl or hard, massive carbonate accumulations, covering lacustrine silt or diatomite. The ancient dune below was bleached to white sand as a consequence of longer inundation. Spatially the layers of lake marl and swamp-ores are always clearly separated although they, as well as the surrounding dune sands, contain small amounts of calcium carbonate. Thin sections of oxide stalks show that the former root channels are coated with a thin layer of calcium carbonate.

The spatial extension of the swamp-ore beds, which are restricted to the contour line of the former paleo-lake shore, proves that the formation of the swamp-ore beds was only possible near the lake shore as well as at shoals with restricted water-depths. Many of the extended paleo-lakes display multiple repetition of the swamp-ore beds from higher relief positions of flat-shore areas towards the paleo-lake basin. They always strictly follow the contour lines (Photo 2).

3.2 Swamp-ore deposits at selected study sites

The spatial distribution and the morphological characteristics of swamp-ores at two study sites are presented in order to demonstrate the variability in morphological characteristics and local peculiarities, which can be used for a genetical and paleo-climatic interpretation.

3.2.1 Sequence of swamp-ores in the shore-area of paleo-lakes of the western Ténéré (Catena 6)

The former flat-shore area of a paleo-lake depression, with a diameter of several km at the study site 6 (Fig. 1), displays banks of swamp ore deposits at different elevations. The oxide-enrichments and the related hydromorphic fossil soils were investigated in profiles 7–10 in catena 6. The sequence of soil profiles stretches from the paleo-lake basin, in a northwesterly direction, to the slope of a flat-shore of a paleo-lake bay (Photo 3, Fig. 2).

The massive bank of swamp ore at profiles 6 and10 (Fig. 2) is several m wide and crossed vertically by former root channels (Photo 3). It denotes a lengthy period of steady lake water level high-stand. A further bank of swamp ore, consisting of single rhizo-concretions from iron oxide, was found to be strongly fragmented in a lower relief position at profile 6,8 (Fig.2). Due to a weaker oxide accumulation, probably due to relatively short phases of level water stands with an overall sinking lake water table, the crust stability was relatively low.

About 150 m apart from the lowermost bank of swamp ore (profile 6,8, Fig. 2) a flat, shield-shaped dune hill, 30 meters in diameter and covered by fragments of swamp ore, rises above the flat bottom of the lake basin (profile 6,7, Fig. 2). The surface of this ancient dune ridge lies about 8 – 10 m below the bank of swamp iron ore at profile 6,8. The substrate of the fossil Gleyic Arenosol, in which the lower part of the swamp ores are embedded, consists of ancient dune sand with silty lacustrine sediments absent. The uppermost decimetres are weakly mottled while the deeper part of the soil profile is white bleached sand. The mottling obviously developed during the final lake phase, when the location of profile 6,7 (Fig. 2) was not permanently inundated but influenced by a high groundwater table. Hence the formation of the lower banks of swamp ore at the paleo-lake shores, as well as the swamp ore formation on top of the shallow dune ridge within the lake basin (which for some period formed a flat) must be assigned to the regression phase of the lake in the course of increasing climatic aridity.

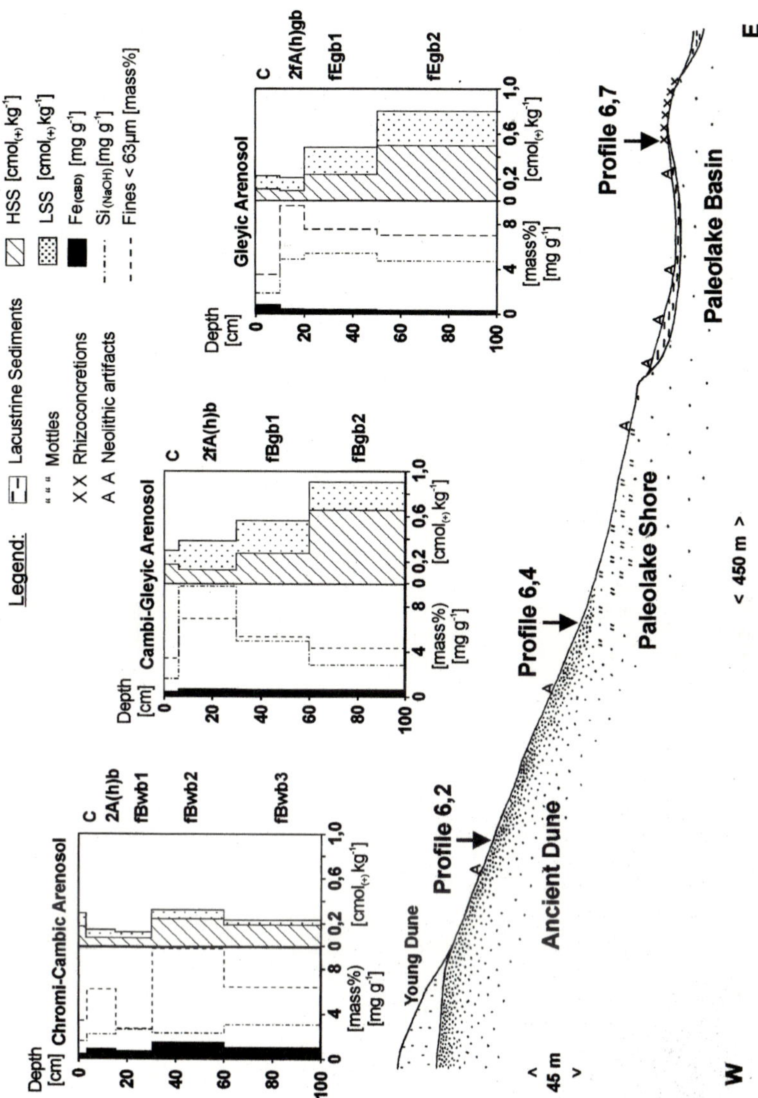

Fig. 2. Typical paleosol catena of Gleyic Arenosols with swamp ore beds on the shore area of a paleolake depression at site 6 in the Western Ténéré (location s. Fig. 1), and depth functions of fines (< 60µm), pedogenic oxides and salts

AA = Neolithic artefacts

XX = Swamp ores and rhizoconcretions from goethite

HSS = Highly soluble salts

LSS = Less soluble salts

$Fe_{(CBD)}$ = Pedogenic ferric iron, extracted by citrate, bicarbonate and dithionite

$Si_{(NaOH)}$ = Amorphous silica, extracted by 0.5 M NaOH

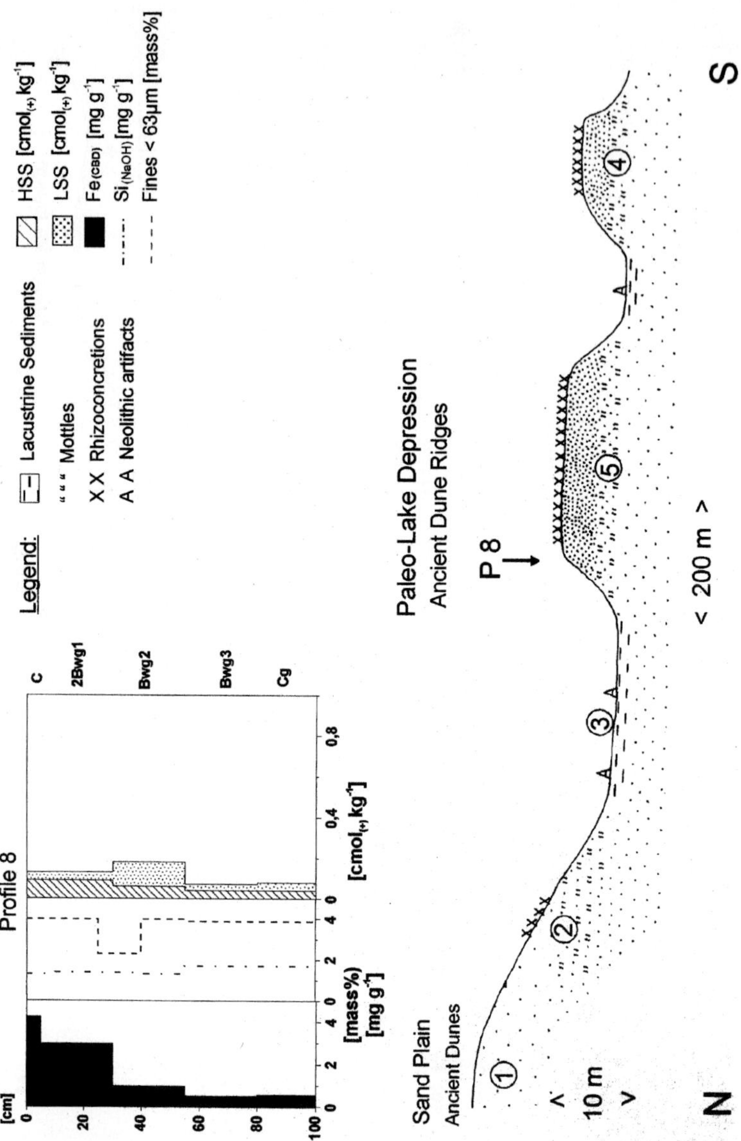

Fig. 3. Typical paleosol catena of Gleyic Arenosols with swamp ore beds on the shore area and on top of ancient dune ridges of a paleolake depression at site 8 in the Eastern Ténéré (location s. Fig. 1), and depth functions of fines (< 60µm) , pedogenic oxides and salts

AA = Neolithic artefacts

XX = Swamp ores and rhizoconcretions from goethite

HSS = Highly soluble salts

LSS = Less soluble salts

$Fe_{(CBD)}$ = Pedogenic ferric iron, extracted by citrate, bicarbonate and dithionite

$Si_{(NaOH)}$ = Amorphous silica, extracted by 0.5 M NaOH

3.2.2 Swamp ores on ancient dune islands within paleo-lakes (Catena 8)

East of the oasis Achegour (site 8, s. Fig. 1) the relief of the Ténéré is a wide flat sand plain, morphologically contoured by single longitudinal dune chains and wide trough valleys within which paleo-lakes existed. Former shore lines are characterized by banks of swamp ores. The bottom of the paleo-lake is covered by gray and black organic silty lacustrine sediments that contain remnants of ferruginized fish bones. At the surface a strew of Neolithic artefacts and fragments of pottery are still preserved, which shows that this paleo-lake was dry and settled during the younger Neolithic. In the middle of the paleo-lake basin, there are up to 8 m high ancient dune ridges that extend like elongate islands in direction of the dominant NE trade wind. The summit area of these ridges is shaped as plateaus, covered by fragments as well as atochthonous banks of swamp ores (Fig. 3).

A soil profile on top of a levelled ancient dune ridge revealed that the iron oxide accumulation of the cauliform swamp ores led to a cementation of the sand down to 30 cm below the recent surface (Photo 4). The dun colouring of the upper part of the swamp ores indicates a higher proportion of Mn-oxides, while the lower part has a yellow brown colour. The surface of the ancient dune ridge slopes, and the plateaus near the edges, are covered with reworked and rounded fragments of cauliform swamp ores (Photo 5). Obviously wave action and later corrasion after the paleo-lake dried out caused a partial destruction of the autochthonous swamp ore banks.

Between and below the banks of swamp ore, the ancient dune sand is weakly cemented and displays a reddish-brown Bw horizon of a Chromi-cambic Arenosol without hydromorphic characteristics. Furthermore, numerous animal burrows are filled with humic sand from a former A horizon. This indicates the presence of a deep groundwater table during the period of soil formation following the decline of the paleo-lake water table during a later phase of the humid period. Or the groundwater table may have been present during a younger, late Holocene humid period with a climate of lower humidity in which the rising lake water table did not reach the plateau level of the ancient dune ridges.

Neolithic tools and pottery are scattered on top of the lacustrine sediments in many paleo-lake basins. This indicates that the lakes later contracted during an arid period and did not reach their full extent again during the mid-Holocene humid period. According to Völkel (1988), a Haplic Arenosol developed on ancient dunes of East Niger during this humid period.

3.3 Total element contents

Because the sand fraction of the ancient dunes consists of more than 95 % quartz (Pfeiffer 1991) the contents of iron oxides and other pedogenic elements of the Ah and Bw horizons of the fossil soils is relatively low (see soil data in Fig. 2 and Table 1). Accordingly, the sources and species of elements, which are enriched in the

Photo 4. A massive swamp ore deposit on top of ancient dune sand, about 20 - 40 cm thick, forming a ridge at the present-day surface due to the deflation of neighbouring lacustrine sediments and dune sands that follow the shore line of the paleo-lake. In the direction of the lake basin (right) several swamp ore deposits occur in different levels with decreasing amounts of oxide concentration (Ténéré site 6, c.f. Fig. 1)

Photo 5. Sequence of massive swamp ore deposits following the flat shore line of the paleo-lake. In the direction of the lake basin (left) several swamp ore deposits occur in different levels with decreasing amounts of oxide concentration (Ténéré site 6, c.f. Fig. 1)

Photo 6. Flat lake shore area of a paleo-lake single standing rhizo-concretions stand out of ancient dune sand (with hydromorphic characteristics) surrounded by fragments of rhizo-concretions (Tchigai mountains, site 11, c.f. Fig. 1)

Photo 7. Autochthonous swamp ore deposit on top of an ancient dune within a paleo-lake depression. The colour of the stem-like rhizoconcretions, grown into one and about 40 cm thick, is dark brown in the upper part and yellowish brown in the lower part (Ténéré, east of Achegour, site 8, cf. Fig. 1). The exposed sand in the background is white, bleached due to gleyification

Photo 8. Fragments of a reworked swamp ore deposit on top an ancient dune island within a paleo-lake depression. The uppermost part of the ancient dune displays a weakly cemented Bw horizon with a reddish brown colour of a Cambic Arenosol, which developed at the end of the lake period or within a younger humid period (Ténéré, site 8, cf. Fig. 1)

swamp ores, are of special interest. Tables 1 and 3 display the total amounts of main and trace elements of different swamp ore deposits. For the purpose of comparison, element concentrations are provided of a fossil Bw horizon of a Chromi-cambic Arenosol, near by the swamp ore bank represented by profile 6,10 (Tables 1 – 4). This paleosol displays average concentrations of several Cambic Arenosols which were investigated in the Sahara of East Niger (s. Felix-Henningsen 2000).

Apart from the main accumulation of Fe a further enrichment of main and trace elements, masked by the over-proportional accumulation of Fe, was revealed through a re-calculation of the element concentrations related to the samples free of goethite. Furthermore, the high loss of ignition of iron oxides influences the concentration of all other elements according to the total amount of accumulated goethite. Thus all element concentrations were calculated free of H_2O^+ and H_2O^-. A direct comparison of the element concen trations between Bw horizons of fossil Cambic Arenosols and the sand matrix of the swamp ores, as well as between swamp ores with a different amount of oxide accumulation, is only possible with element concentrations calculated free of goethite and H_2O (Tables 2 and 4).

Table 1. Contents of main element oxides (mass-%, x-ray fluorescence analysis) of swamp iron ores and a neighbouring paleosol (fossil Chromi-Cambic Arenosol) from ancient Saharan desert dunes in East Niger; sites cf. Fig. 1; "T" = traces below detection limit; Totals are < 100 % due to loss by ignition

Bulk samples. mass-%	P 5: Swamp ore	P 6.10: Swamp ore	P7: Swamp ore	P8: Swamp ore. upper part	P8: Swamp ore. lower part	P11: Swamp ore	Paleosol bBw horizon	Paleosol. C horizon
Bulk density	2.33	2.46	2.29	2.72	2.29	2.42	1.64	1.64
SiO_2	45.6	66.3	72.82	65.72	79.59	33.71	95.58	94.89
Al_2O_3	0.80	1.12	1.17	1.06	1.53	0.69	1.99	1.76
Fe_2O_3	44.81	26.74	19.87	26.97	14.93	55.30	0.47	0.40
MnO	0.07	T	T	0.06	T	0.90	T	T
MgO	T	T	T	T	0.11	0.14	T	T
CaO	0.24	0.28	0.20	0.40	0.11	0.34	0.09	0.13
Na_2O	T	T	T	T	T	T	T	T
K_2O	0.19	0.14	T	0.14	0.39	0.14	0.43	0.39
TiO_2	0.07	0.10	0.12	0.12	0.14	0.11	0.15	0.15
P_2O_5	T	0.15	0.10	0.06	0.03	0.15	T	T

Table 2. Contents of main element oxides (mass-%, x-ray fluorescence analysis), calculated Fe and H_2O free, of swamp iron ores and a neighbouring paleosol (fossil Chromi-Cambic Arenosol) from ancient Saharan desert dunes in East Niger; sites cf. Fig. 1; "T" = traces below detection limit

Bulk samples. mass-%	P 5: Swamp ore	P 6.10: Swamp ore	P7: Swamp ore	P8: Swamp ore. upper part	P8: Swamp ore. lower part	P11: Swamp ore	Paleosol bBw horizon	Paleosol. C horizon
Bulk density								
SiO_2	97.14	95.78	97.41	96.69	97.07	93.35	96.03	95.27
Al_2O_3	1.71	1.54	1.65	1,56	1.87	1.91	2.00	1.77
Fe_2O_3								
MnO	0.15	T	T	0.09	T	2.49	T	T
MgO	T	T	T	T	0.13	0.31	T	T
CaO	0.51	0.26	0.41	0.59	0.13	0.94	0.09	0.13
Na_2O	T	T	T	T	T	T	T	T
K_2O	0.41	T	T	0.21	0.48	0.39	0.43	0.39
TiO_2	0.15	0.13	0.14	0.18	0.17	0.30	0.15	0.15
P_2O_5	T	0.13	0.22	0.09	0.04	0.42	T	T

The bulk densities of the swamp ores vary between 2.3 and 2.7 g cm^{-3} and represent clearly lower values than the specific weight of goethite (4.8 g cm^{-3}) due to the internal porosity and the proportion of quartz sand. The accumulation of iron oxides varies between 20 and 55 % mass goethite. Small amounts of Mn, Mg and P, all missing in the Bw horizons of the terrestrial soils, .were accumulated together with iron. The swamp ore of site 8 (Photo 4) shows an accumulation of Mn only in the dun coloured upper part of the cauliform concretions, which indicates a redox gradient with increasing potentials from bottom to top. The relatively high concentrations of Ca, compared to the Bw horizon, can be attributed to the precipitation of carbonates, forming coatings or complete fillings of pores within the ores. The contents of K of the swamp ores are variable and mostly lower as in the Bw horizons, indicating low concentrations and mobility of potassium in the water of the paleo-lake as well as a possible uptake by the swamp vegetation. On the other hand the concentration of Ti (Table 2), an element which is nearly completely immobile and therefore not affected by redistribution processes, is very similar in the Bw horizons and the sand-matrix of the swamp ores, indicating the petrologic similarity of the sands. Also the Zr concentration, which mainly reflects the contents of the heavy mineral zircon, indicates the narrow similarity between the ancient dune sand of the terrestrial soils and the swamp ores.

Among the trace elements the heavy metals Co, Cr, Cu, Ni, V and Zn show an accumulation in the swamp ores as compared to the Bw horizons (Tables 3 and 4). Ba, Rb and Sr, mainly bound in carbonates and silicates, display a higher variability without a significant trend due to the concentration of the host minerals.

4 Genesis and paleo-ecological interpretation

4.1 Processes of formation

The spatial distribution, and consequently the formation of swamp iron ores, was systematically bound to shallow water areas at flat shores and flats within Holocene paleo-lakes. The swamp ores display massive concentrations of iron oxides that were precipitated by oxidation of Fe^{2+} ions dissolved in the lake water. Due to morphological characteristics, the concentration of the iron oxides occurred in the upper root zone near the lake bottom, as they display cauliform rhizo-concretions with root channels and side branches. They consist mainly of goethite without Al substitution, since dissolution and mobility of aluminium ions in the lake water was not possible due to the high pH as a consequence of the carbonate contents of the lacustrine sediments. A proportion of lepidocrocite in the outer zone of the cauliform swamp ores was also identified by Schwertmann and Taylor (1989) in rhizo-concretions from water-logged soils and interpreted as a consequence of differences in the CO_2 partial pressure between zones near (high pCO_2) and distant to the root channel (lower pCO_2). A steep gradient in pCO_2 over a short distance can result from CO_2 excretion of roots or by the low solubility of CO_2 in the higher temperatures of the shallow lake water. P, Mn, Co, Cr, Ni , V and Zn, which also were mobilized in the lake water by reduction, accumulated by oxidation (Mn) or adsorption to the freshly precipitated iron oxides. The precipitation was mainly bound to the surface of vegetation organs of the transition zone between lower stem and upper roots, as hardly any isolated concretions - apart from some mottles of iron oxide - occur in horizons below or alongside the banks of swamp ores. The oxides coated the vegetation organs and precipitated within pore spaces of the surrounding sediment. Therefore they often show a concentric lamination around the former root channels and branches of side roots.

From the paleo-lake shore towards the basin the morphology and occurrence of the swamp ores shows a clear zonation. At shallow lake shores, or in flat areas within the paleo-lake basin with shallow water, conditions for the growth of water vegetation was favourable, while this was less likely at steep lake shores with deep water. With increasing water depth in the direction of the lake basin, both the density of the vegetation and the thickness of single plants decreased. Therefore single standing and thinner rhizo-concretions were formed. Another explanation for the formation of singular rhizo-concretions could have been a rapid decline in water depth due to increasing aridity. The retention period of the water vegetation at depth, which was favourable for the formation of swamp ores, was thus too short.

Nevertheless, the relatively sharp limitation of the swamp ore banks against the bleached lacustrine sediments and ancient dune sands of the deeper lake basin indicates a pronounced lateral redox gradient within the lake water and the sediment of the paleo-lakes. The contents of calcium carbonate of the lacustrine sediments and the deposits of lake carbonates prove that the lake water was buffered by bicarbonate and therefore should have had a neutral to weakly alkaline pH. Therefore, according to the Eh-pH stability conditions (see Hem 1972, Skinner and Fitzpatrick 1992), the redox gradient was primarily influenced by the oxygen concentration of the lake water. A reducing environment existed in deeper water

Table 3. Contents of trace element oxides (mg kg^{-1}, x-ray fluorescence analysis) of swamp iron ores and a neighbouring paleosol (fossil Chromi-CambicArenosol) from ancient Saharan desert dunes in East Niger ancient dune ; sites cf. Fig. 1; "T" = traces below detection limit

Bulk samples. mass-%	P 5: Swamp ore	P 6.10: Swamp ore	P7: Swamp ore	P8: Swamp ore. upper part	P8: Swamp ore. lower part	P11: Swamp ore	Paleosol bBw horizon	Paleosol. C horizon
Ba	198	105	47	132	191	56	161	140
Co	47	22	28	40	28	46	T	T
Cr	11	29	24	11	20	16	T	10
Cu	14	16	T	T	T	T	T	T
Ni	60	25	T	34	T	25	T	9
Pb	5	8	21	26	26	5	10	11
Rb	T	5	10	10	9	4	18	5
Sr	10	22	18	20	28	6	38	30
V	79	149	37	28	T	22	T	T
Zn	22	26	30	24	14	19	15	17
Zr	177	229	230	234	256	320	348	346

Table 4. Contents of trace elements (mg kg^{-1}, x-ray fluorescence analysis), calculated Fe and H_2O free, of swamp iron ores and a neighbouring paleosol (fossil Chromi-Cambic Arenosol) from ancient Saharan desert dunes in East Niger; sites cf. Fig. 1; "T" = traces below detection limit

Bulk samples. mass-%	P 5: Swamp ore	P 6.10: Swamp ore	P7: Swamp ore	P8: Swamp ore. upper part	P8: Swamp ore. lower part	P11: Swamp ore	Paleosol bBw horizon	Paleosol. C horizon
Ba	422	62	154	194	233	155	162	141
Co	100	37	32	59	34	127	T	T
Cr	23	32	43	16	24	44	T	10
Cu	30	T	24	T	T	T	T	T
Ni	128	T	37	50	T	69	T	9
Pb	11	28	12	38	32	14	10	11
Rb	T	13	7	15	11	11	18	5
Sr	21	24	32	29	34	17	38	30
V	168	49	219	41	T	61	T	T
Zn	47	39	38	35	17	53	15	17
Zr	377	303	337	344	312	886	348	346

zones and at the lake bottom as a consequence of a low oxygen content due to high water temperatures (s. Schwoerbel 1987, Lampert and Sommer 1993) and the decomposition of organic material derived from the sedimentation of dead algae and water animals. Organic layers within the lacustrine sediments on top of the bleached ancient dune sand still occur today in areas protected from deflation. The reducing environment caused the dissolution of the iron oxides of the ancient dunes and dusts, which were deposited in the lake basin, by the frequent dust storms that occur in the Sahelian environment. The reduced heavy metal ions as well as dissolved phosphorous diffused from the sediment into the lake water along a concentration gradient (s. Ponnamperuma 1972). This caused deep bleaching (many meters) of the lacustrine sediments and the underlying ancient dunes. The greater part of the dissolved ions moved by diffusion and mass flow within the reducing zones of the lake water in the direction of the vegetated flat shores and flats with shallow water. Here the iron and manganese ions were precipitated by oxidation, incorporating other heavy metals and phosphorus by adsoption. Massive accumulation over a long period of time led to the formation of the banks of swamp iron ores. Furthermore, Fe and Mn oxides were accumulated in mottles and concretions within the uppermost, partly oxidized Bwg horizons of the Gleyic Arenosols distributed along the paleo-lake shores. Due to the oxidation processes the redox gradient in the lake water continued, a precondition for the diffusion and accumulation of the ions. But high redox potentials, which could have led to a precipitation of iron and manganese oxides, are not typical in such conditions. Vegetation provides litter, which decomposes microbially, causing a consumption of oxygen. Furthermore, the shallow water of flat shore areas became strongly heated. This caused a loss of oxygen as a consequence of a low physical gas dissolution and the promotion of microbial activity. According to Lampert and Sommer (1993) the oxygen concentration of polymictic tropical lakes varies in a daily course along with solar radiation and, as a consequence, the heated lake zones are nearly free of oxygen. However, wave activity near the lake shore could have supported the intake of oxygen. But this should have led to the formation of oxide sludge on the lake bottom only under continuously high redox potentials, which cannot be expected according to the reasons mentioned above, and not to an accumulation around stems and roots of the vegetation within the upper soil horizon. This could only have occurred by diffusion of dissolved reduced ions from the free water into the sediment, which, with respect to the release of elements, was the reverse process.

Only the specific physiological characteristics of the water and swamp vegetation can provide a plausible explanation for the oxidative accumulation of oxides.

Reeds, rushes, papyrus, marsh vegetation and rice, and other swamp species, bound to photosynthesis with a sprout growing above the water table, but rooted in permanently reduced soils, are provided with an aerenchyma which enables the oxygen supply of the roots (Trolldenier 1977, Ando et al. 1983, Grosse and Wilhelm 1984). The oxygen surplus not consumed by respiration (alpha naphtylamin oxidation) diffuses into the intercellular spaces and through the epidermis in the adjacent soil pores. This causes high redox potentials within the epidermis, on the root surface and in the soil surrounding the roots and prevents toxic reduced chemical compounds, such as H_2S, Fe^{2+} sulfides or organic acids, from ingressing the roots (e.g. Tanaka et al. 1966, Howeler 1973). The transport of oxygen from the sprout to the roots is independent of light exposure and therefore enables root respiration even in darkness. The rates of oxygen release through the root epidermis increases with decreasing temperature because the amount used for respiration processes decreases. Roots of rice, which grow in water rich in reduced metal ions, show oxide coatings on the root surfaces. They consist mainly of Fe oxides, partly of low crystalline lepidocrocite. Mn oxides, however, are only precipitated under conditions of a relatively high Mn^{2+} concentration (Bacha and Hosner 1977), which is congruent with the mineralogical composition of the swamp ores. Trolldenier (1988) found that bacteria participated in the oxidation of iron on the root surface. The coating by iron oxides of the root surface of swamp vegetation, growing in water rich in dissolved Fe^{2+}, can lead to iron toxicity caused by inhibition of the nutrient uptake, mainly of P, Zn, K, Ca, Mg, which are adsorbed to the freshly precipitated iron oxides (e.g. Tanaka et al. 1966, Howeler 1973). These findings suggest the formation of the swamp ore banks along the shore lines of the Ténéré paleo-lakes as a consequence of the oxidizing properties of the roots of the swamp vegetation.

A rough estimated mass balance provides an impression of the consequence of these processes.

Unweathered ancient dune sands (C horizon) display free iron oxides in the range of 0.5 – 1 mg Fe_{CBD} g^{-1}. The concentration amounts to 2 mg Fe_{CBD} g^{-1} (s. Fig. 2) in bBw horizons of Chromi-cambic Arenosols, which developed during the early Holocene humid period on the ancient dunes. The pedogenic oxides coat the quartz grains of the sand fraction, which causes the brownish yellow colour of the ancient dune sands. From 1 m^3 of ancient dune sand with a bulk density of 1.6 g cm^{-3}, an amount of 800 – 1,600 g Fe was sufficient to form 470 – 940 cm^3 of swamp ore from goethite with a bulk density of 2.7 g cm^{-3}. Therefore 1 m^2 bank of massive swamp ore, 30 cm thick, could only have developed by accumulation of iron from 600 – 1,200 m^3 of ancient dune sand. At many paleo-lake shores the spatial extension of such banks of swamp ores exceeds many tens of meters (see Photo 3) and they can occur with several repetitions at different elevations. The calculation highlights the enormous redistribution of elements within the paleo-lake basins, proven by the extension and depth of the white, bleached ancient dune sands and lacustrine sediments.

As a further source of accumulated elements, dust deposition into the paleo-lakes must be considered. During the humid periods the existence of a Sahelian ecosystem can be assumed (e.g. Baumhauer et al. 1989). Hence, dust deposition on soils and paleo-lakes derived from long-distance transport with the monsoon Harmattan, which still occurs in the Sahelian zone (Stahr et al. 1994, Herrmann 1996). Remnants of lacustrine sediments, rich in silt that occur within paleo-lake depressions protected from deflation, prove the eolian deposition of dust. According to Herrmann (1996) recent dust deposition in the Sahelian zone consists of > 70 % mass of silt and contains carbonates. This explains the formation of lake carbonates because the ancient dune sands and the underlying Cretaceous sand stones are normally free of calcium carbonate. Along with the dust, other basic cations as well as Fe, P and heavy metals, were deposited.

The possible precipitation and accumulation of oxides due to ground-water springs into the lake water at lake shores is an alternative, though less probable hypothesis. Fe^{2+} concentrations of the ground water bodies of the ancient dunes should have been very low. Soil organic matter was almost completely mineralized within the highly permeable, aerated and well buffered topsoil horizons. Therefore the unsaturated zone of the Cambic Arenosols shows no characteristics of redox processes, which would indicate the mobilization of Fe-Oxides. Reduction processes were exclusively bound to the inundated soils of the lake basin where the release of reducing agents from organic matter was possible. Furthermore, groundwater springs should preferentially have occurred at steep shores rather than at vegetated flat shores. If the redox potentials of the near-shore lake water would have enabled oxidation of Fe^{2+} ions, which in vegetated shallow water zones (according to the arguments presented above) was nearly impossible, the formation of iron oxide sludge - and therefore platy crusts of iron oxide on top of the lacustrine sediments - would have resulted.

In addition, the genesis of swamp ores differs from that of “root tubules”, “pipestems” or rhizo-concretions, which frequently occur in soils with stagnic properties or partly in ferrallitc soils (e.g. Schwertmann and Taylor 1989, Fitzpatrick 1988, Schwertmann and Fitzpatrick 1992). As a consequence of water logging, Fe^{2+} ions from the reduction of iron oxides of the Bwg horizons reprecipitated as oxides at the walls of macro-pores and aerated empty root channels.

4.2 Paleo-ecological indication

Apart from being found in paleo-lake sediments and paleosols on ancient dunes, swamp iron ores are indicators of ecosystem structures and the geomorphological and paleo-climatic processes of landscape development during humid periods. They can establish the spatial extension of paleo-lakes and the distribution of former swamp vegetation. With variable morphology, from massive banks to thin, single standing cauliform rhizo-concretions, they are distributed in large areas of the central and southern Sahara of East Niger. Differences in the spatial density of the occurrence of swamp ores obviously exist between the Ténéré and the Tchigai mountains on the one hand, and the Grand Erg de Bilma on the other. Compared to the flat undulating relief of the Ténéré and the extensive basins of the Tchigai mountains, the ancient dune relief of Grand Erg de Bilma is more pronounced. The embedded paleo-lake basins had steeper shores with limited shallow water areas; preconditions for the formation of swamp ores. Thus they occur locally in the Erg of Bilma as small banks in single flat bays.

Considering the paleo-climatic phases of the early Holocene humid period, the sequences of swamp ores at flat shores of paleo-lakes are of importance. Each bank of swamp ore at different altitudes indicates a long period with a stable lake water table. The sequence of swamp ores from the orographically highest bank to the lower banks, however, indicates a declining water table due to increasing aridity. Obviously this trend was interrupted several times by phases of relative stable humidity, and therefore a stable position of the shore line led to the accumulation of banks of swamp ores. The duration of these climatic phases was sufficient such that a Bw horizon was formed in ancient dunes, surrounding the oldest banks of swamp ores in the highest landscape positions.

During the following middle Holocene arid period the paleo-lakes dried out at different periods, according to the depth of the paleo-lake depressions and the extension of local ground water aquifers. During the following middle Holocene humid period many of the paleo-lake bottoms were settled by man, while other paleo-lakes still existed (Baumhauer 1993).

5 Acknowledgements

Fieldwork in the Sahara was only made possible by the professional organisation and preparation of the expedition by R. Baumhauer (University of Trier), D. Busche and B. Sponholz (University of Würzburg) and helpfulness of A. Grote (University of Münster). I am grateful for their active support in the field and for many fruitful interdisciplinary discussions. Their assistance, as well as the generous funding of the project by the Deutsche Forschungsgemeinschaft (DFG), is gratefully acknowledged.

6 References

Ando, T., Yoshida , S. and Nishiyama, I. (1983): Nature of oxidizing power of rice roots. - Plant and Soil 72: 57- 71.

Armstrong, W. (1967): The oxidizing activity of roots in waterlogged soils. - Physiol. Plant, 20: 920- 926.

Bacha, R.E. & Hossner, L.R. (1977): Characteristics of coatings formed on rice roots as affected by iron and manganese additions. - Soil Sci. Soc. Am. J. 41:931- 935.

Baumhauer, R. (1991): Paleo-lakes of the south central Sahara - problems of paleoclimatical interpretation. - Hydrobiologia, 214: 347- 357

Baumhauer, R. (1993): Probleme der paläoökologischen Interpretation limnischer Akkumulationen im Ténéré, NE-Niger. - Trierer Geogr. Studien, 9: 33 - 49

Baumhauer, R., Busche, D. und Sponholz ,B. (1989): Reliefgeschichte und Paläoklima des saharischen Ost-Niger. - Geograph. Rundschau, 41: 493 - 499.

Felix-Henningsen, P. (1992): Merkmale, Verbreitung und klimazonale Ausprägung frühholozäner Feuchtzeitböden in der Ténéré, Ostniger. - Würzburger Geogr. Arbeiten, 84: 97 - 129.

Felix-Henningsen, P. (2000): Paleosols on Pleistocene dunes as indicators of paleomonsoon events in the Sahara of East Niger. Catena 41: 43 – 60.

Fitzpatrick, R.W. (1988): Iron compounds as indicators of pedogenic processes: Examples from the southern hemisphere. - In: J.W. Stucki, B.A. Goodman & U. Schwertmann (Eds.): Iron in soils and clay minerals, S. 351 - 396, NATO ASI Vol. 217, Reidel, Dordrecht.

Grosse, H. & Wilhelm, A. (1984): Sauerstoffversorgung bei Wasserpflanzen (Druckventilation). - Z. Biologie in unserer Zeit, 14: 28 - 31

Grunert, J. (1988): Klima- und Landschaftsentwicklung in Ost-Niger während des Jungpleistozäns und Holozäns. - Würzburger Geogr. Arb. 69: 289 - 304

Grunert, J. Baumhauer, R. & Völkel, J. (1991): Lacustrine sediments and Holocene climates in the southern Sahara: the example of palaeolakes in the Grand Erg of Bilma (Zoo Baba and Dibella, Eastern Niger. - Journal of African Earth Sciences, 11: 133 - 146.

Hem, J.D. (1972): Chemical factors that influence the availability of iron and manganese in aqueous systems. - Geol. Soc. Am. Bull., 83: 443 - 450.

Herrmann, L. (1996): Staubdeposition auf Böden Westafrikas. - Hohenheimer Bodenkundliche Hefte, 36: 1 - 239.

Howeler, R.H. (1973): Iron-induced oranging disease of rice in relation to physical-chemical changes in a flooded oxisol. - Soil Sci. Soc. Am. Proc. 37: 898 - 903.

Kröpelin, S. (1993): Zur Rekonstruktion der spätquartären Umwelt am Unteren Wadi Howar (Südöstliche Sahara/ NW-Sudan). - Berliner Geogr. Abh. 54: 1 - 293.

Lampert, W. & Sommer, U. (1993): Limnoökologie. - 440 S., Thieme, Stuttgart, New York.

Maley, J. (1981): Etudes palynologiques dans le bassin du Tschad et paleoclimatologie de l'Afrique nord-tropicale de 30.000 ans à l'epoque actuelle. - Travaux et documents de l'O.R.S.T.O.M 129 : 1 - 586.

Pfeiffer, L. (1991): Schwermineralanalysen an Dünensanden aus Trockengebieten mit Beispielen aus Südsahara, Sahel und Sudan sowie Namib und der Taklamatan. - Bonner Geogr. Abh. 83: 216 S.

Ponnamperuma, F.N. (1972): The chemistry of submerged soils. - Adv. Agron. 24: 29 - 96.

Schlichting, E., Blume, H.-P. & Stahr, K. (1995): Bodenkundliches Praktikum. - 2. Aufl., 295 S., Blackwell, Berlin, Wien.

Schwarz, T. (1992): Produkte und Prozesse exogener Fe-Akkumulation: Eisenoolithe und lateritische Eisenkrusten im Sudan. - Berliner Geowiss. Abh. A 142, 186 S., Berlin.

Schwertmann, U. & Taylor, R.M. (1989): Iron oxides. In: Soil Science Society of America (Eds.): Minerals in soil environments. Chapter 8: 379 - 438, Madison.

Schwertmann, U. & Fitzpatrick, R.W. (1992): Iron minerals in surface environments. - In: H.C.W. Skinner & R.W. Fitzpatrick (Eds.): Biomineralization processes of iron and manganese. - Catena Suppl. 21: 7 - 30, Cremlingen.

Schwoerbel, J. (1987): Einführung in die Limnologie. - 6. Aufl., 269 S., UTB 31, Fischer, Stuttgart.

Skinner, H.C.W. & Fitzpatrick, R.W. (1992): Iron and manganese biomineralization. - In: Skinner, H.C.W. & Fitzpatrick, R.W. (Eds.): Biomineralization. - Catena Suppl. 21: 1 - 6, Cremlingen

Stahr, K. Herrmann, L. & Jahn, R. (1994): Long distance dust transport in the Sudano-Sahelian zone and the consequences for soil properties. - In : Buerkert, B., Allison, B.R. & Oppen, M. von (Eds..): Proceedings of the International Symposium "Wind erosion in West Africa": 23 - 33, Hohenheim.

Tanaka, A., Loe, R. & Navasero, S.A. (1966): Some mechanisms involved in the development of iron toxicity symptoms in the rice plant. - Soil Sci. Plant Nutr. 12: 158 - 164.

Trolldenier, G. (1977): Mineral nutrition and reduction processes in the rhizosphere of rice. - Plant and Soil 47: 193 - 202

Trolldenier, G. (1988): Visualisation of oxidizing power of rice roots and of possible participation of bacteria in iron deposition. - Z. Pflanzenernähr. Bodenk. 151: 117 - 121.

Völkel, J. (1988): Zum jungquartären Klimawandel im saharischen und sahelischen Ost-Niger aus bodenkundlicher Sicht. - Würzb. Geogr. Arb. 69: 255 - 276

Völkel, J. (1989): Geomorphologische und pedologische Untersuchungen zum jungquartären Klimawandel in den Dünengebieten Ost-Nigers (Südsahara und Sahel). - Bonner Geogr. Abh. 79: 258 S..

Williams, M. A. J., Abell, P. I. & Sparks, B. W. (1987): Quaternary landforms, sediments, depositional environments and gastropod ratios at Adrar Bous, Ténéré desert of Niger, South-Central Sahara. - In: Frostick, L.E. & Reid, I. (Eds.): Desert sediments: ancient and modern. - Geological Soc. spec. publ. 35, Blackwell, Oxford.

Fulgurites as palaeoclimatic indicators — the proof of fulgurite fragments in sand samples

Barbara Sponholz

Department of Geography, University of Würzburg,
Am Hubland, D-97074 Würzburg, Germany
barbara.sponholz@mail.uni-wuerzburg.de

Abstract

Fulgurites are formed by lightning strikes to sandy ground. The paper describes the occurence of fulgurites in the southern Sahara (eastern Niger) and their palaeoclimatic relevance. All the studied fulgurite fragments were found near to palaeolake sediments in midslope positions of interdune depressions. The mineralogical composition (lechatelierite, cristobalite, chalcedony, opal) of the fulgurites is related to the palaeo-environmental conditions of the semi-(arid) regions and to the melting conditions during the fulgurite forming lightning strike to the ground.

Introduction

Fulgurites (latin: "fulgur" = lightning) are exclusively formed by lightning strikes to the ground. Therefore fossil fulgurites indicate former lightning and thunderstorm activity as well as thunderstorm-related rainfall. Their general value for palaeoclimatic reconstruction has been pointed out by SPONHOLZ et al. (1993).

In areas where lightning hits mostly quartz sands (e.g. in dunes, river terraces, etc...), melting of the quartz sands will take place at very high temperatures (up to 3,000 K; after FELDMANN 1988) followed by immediate cooling. These processes cause the melted quartz to transform itself into an amorphous substance. This newly formed mineral product, the socalled "lechatelierit", is one of the natural glasses beneath volcanic glasses, tektites and impactites. Also cristobalite may occur as one of the high-temperature modifications of SiO_2 in the external parts of the fulgurites.

As melting occurs along the lightning path through the ground, the fulgurites are more or less cylindrical. Their total thickness is up to several centimeters, the glassy wall around the central tubular void — the former lightning channel — being up to a maximum of some millimeters thick (Fig. 1). This form makes fulgurites very fragile and susceptible to mechanical stress.

Palaeoclimatic interpretation

A very important concentration of fulgurite fragments up to 30 cm long found in the central and the southern Sahara of Eastern Niger (south of 18°N) was interpreted by SPONHOLZ et al. (1993) for palaeoclimatic purposes. All studied fragments have been found close to Holocene palaeolake sediments in midslope positions of interdune depressions. This is explained by the presence of dominant electrical fields in the midslope position during thunderstorm events. The reason for the strong electrical field formation is the hydrological situation during ful-

Fig. 1. Typical fulgurite fragments formed from Saharan dune sands. Because of the thin glassy lechatelierite wall around the inner void of the cylinder they are very susceptible to any mechanical stress. Fluvial action destroys them very rapidly, as well as trampling and strong corrosion

gurite formation: the waterlevel of the lake-filled depressions is linked to the groundwater level inside the surrounding dunes. Between the water-saturated dune sands at the base, and the overlying dry dune sands, the electric tension is strong enough to direct the lightning strikes to the midslope position. The geomorphological, sedimentological and archaeological characteristics (see SPONHOLZ et al. 1993) of the studied fulgurite sites indicate a main period of fulgurite formation during the middle and the upper Holocene, becoming more recent from north to south. This is associated with the development of the (palaeo-)monsoon during the Holocene and the progressive southward movement of the northermost monsoon‘s thunderstorm limit. The preservation of the fulgurites around the palaeolakes/interdune depressions, however, is restricted by the fact that in such positions an important destruction of fossil fulgurites is probable because of trampling by both man and animals near these former waterpoints.

Another reason for considerable mechanical fulgurite destruction in the Sahara is corrasion by the strong winds, e.g. in the southern foreland of the Tibesti Mountains west of Faya (Tchad).

In order to support the palaeoclimatic interpretation of the fulgurite distribution, mapping and statistical analyses were carried out. As many of the fulgurites should have been completely fragmentized by the above mentioned mechanical stresses since their formation in the Mid-Holocene, even the evidence of very small fragments (sand size) is helpful.

Related to the total volume of a dune body, the share of fulgurites does not exceed several %o. Therefore, the presence of sand size fulgurite fragments in several samples taken from adjacent sampling sites gives the sure information of former fulgurite presence at the same place. Even a short-distance transport of the fulgurite fragments would minimize the very small fulgurite share of the complete dune body so much that adjacent sampling sites would not necessarily contain fulgurite fragments any more.

So how can we get proof of sand size fulgurite fragments?

Analytical studies

The fulgurites formed in quartz sands are composed of natural quartz glass, "lechatelierite", and sometimes of cortex grains of cristobalite. Both silica variations are characterized by their crystallographic structure or "non-structure", respectively: The quartz crystal with regular extinction under polarised light changes to a similar image in cristobalite, but with "bag-like" disturbances all over the crystal (PICHLER & SCHMITT-RIEHGRAF 1987). The most important part of the fulgurites, however, is composed of lechatelierite. This mineral does not show any organised crystallographic structure under polarised light, but it looks exactly like the sample bearing glass slide, i.e. a perfect isotropic substance. It shows no extinction and no crystallographic orientation. This makes the lechatelierite clearly different from the original quartz sands. In addition, from amorphous to microcrystalline silica the lechatelierite is easy to distinguish. Opal and chalcedony occur in the foreland of the Tibesti Mountains by volcanic influence or in other parts of the southern Sahara as surface formation, e.g. in plant silicifications. While these types of „amorphous“ to microcrystalline silica contain a certain amount of water and display at least a minimal stage of crystallographic orientation, the lechatelierite does not have these properties.

Separation of both minerals — amorphous silica and lechatelierite — by the specific mineral weight is not possible, however, because the amorphous varieties are too heterogeneous in their water content and both may have impurities.

In order to identify the above mentioned minerals, thin sections of several samples were made: sand sized grains (100 — 500 μm grain diameter) were placed on glass slides and fixed by resin. Afterwards they were prepared as thin sections and have been counted for their mineral content under polarised light. The samples consist of

1. a fulgurite fragmentised in the Specks mill (sample from the southern Sahara/Niger) (Fig. 2),
2. a sand sample of the southern Sahara taken several tens of km from the next known fulgurite site (sample from the Bahr el Ghazal area/Tchad) (Fig. 3),
3. several sand samples taken from fulgurite sites in interdune depressions (western Sahara/Mauritania)
4. an opal fragmentised in the Specks mill (Falaise d'Angamma/Tchad) (Fig. 4)

The counting of the grain types per sample indicated:

1. almost pure lechatelierite, no cristobalite in this preparation, some quartz grains (non transformed distal parts of fulgurites or impurities)
2. no single grain of lechatelierite or cristobalite; only quartz sands and some impurities (non-specified)
3. two grains of lechatelierite (about 1 % of all counted grains) in a matrix of "normal" quartz grains and some impurities,
4. several types of opal in a variety of crystallographic structure and of optical extinctions.

This demonstrates that thin section analysis can help to distinguish fulgurite bearing and non-fulgurite bearing sand samples.

The preparation of thin sections of sand samples (after "natural" sampling in the field or artificial fragmentation of defined minerals) does not present a difficulty. This method provides a reliable and easy way to prove (former) fulgurite existence e.g. in dune areas or in fluvial (quartz rich) sediments. In regions where fulgurites have formed under palaeoclimatic conditions, at least several grains of lechatelierite are present in the sand samples even if the larger fulgurite fragments have been completely destroyed.

With the increasing fragmentation the regional information gets somewhat less precise because of a certain influence of sand transport. On the other hand, this method allows a denser network of mineralogical analyses on fulgurite distribution over a certain area than only macroscopic fulgurite mapping does. By the help of further studies upon this problem the knowledge about Holocene fulgurite formation sites will be more precise.

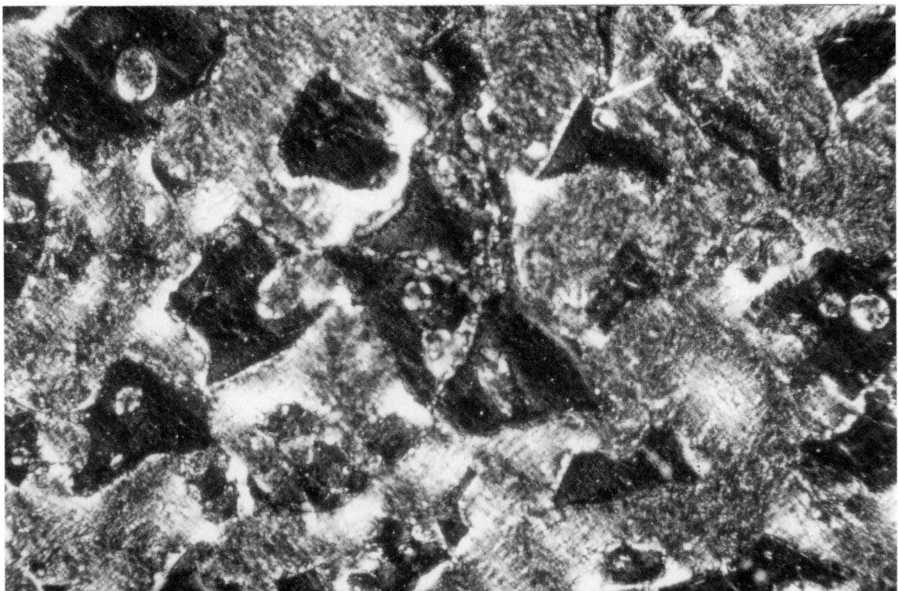

Fig. 2. Fulgurite/lechatelierite, artificially fragmentised in the Specks mill (thin section under polarised light). The regular dark spots are lechatelierite, the strucural matrix is formed by the resin preparation

Fig. 3. Quartz sand from the Bahr el Ghazal/Tchad

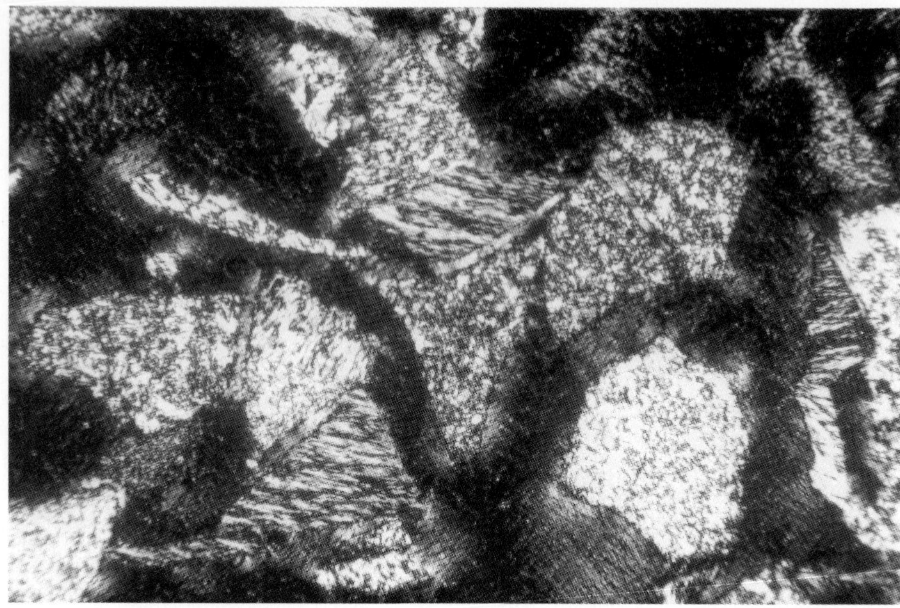

Fig. 4. Opal and chalcedony, partially quartzine, artificially fragmentised in the Specks mill (thin section under polarised light)

References

Feldmann, V. 1988. Comparative Characteristics of Impactite, Tektite and Fulgurite Glasses. *In Konta, J. (editor), International Conference on Natural Glasses, Prague* **1987**, 215-220.

Pichler, H. & Schmitt-Riehgraf, C., 1987. Gesteinsbildende Minerale im Dünnschliff. *Enke, Stuttgart, 230 p.*

Sponholz, B., Baumhauer, R. & Felix-Henningsen, P., 1993. Fulgurites in the southern Central Sahara, Republic of Niger, and their palaeoenvironmental significance. *The Holocene*, **3,2**, 97-104.

Alluvial loess in the Central Sinai: Occurrence, origin, and palaeoclimatological consideration

Rögner, Konrad[1)]; Knabe, Katharina[2)]; Roscher, Bernd[2)]; Smykatz-Kloss, Werner[2)] & Zöller, Ludwig[3)]

[1)] Institute for Geography, LMU Munich
[2)] Institute for Mineralogy and Geochemistry, University of Karlsruhe
[3)] Geographical Institute, University of Bayreuth

Abstract

The layered silts of Wadi Feiran and its tributaries (Central Sinai, Egypt) form profiles up to 50 m in height. They have been the subject of numerous geoscientific studies, but their formation and origin is still under discussion (i.e. lacustrine? fluvial? glaciation? true loess? river terraces?). The investigations of the authors confirm the silts to be alluvial loess. Due to Miocene foraminifera embedded in some of the silts around the oasis of Feiran, the origin of the aeolian material can be traced back to the Gulf of Suez. After sedimenting on the slopes of the wadis (Feiran, Es Sheikh, Solaf) the silts were later washed out by the rain, transported by a meandering river and sedimented as "overbank fines" and "crevasse splays" next to coarser material. At some locations the sedimentation took place in a swamp-like environment. Thermoluminescence dating revealed the time of sedimentation (as overbank fines etc.) being between 27 and 12 ka old. Based on geological and geochemical data the palaeoclimatologic development of the region is discussed.

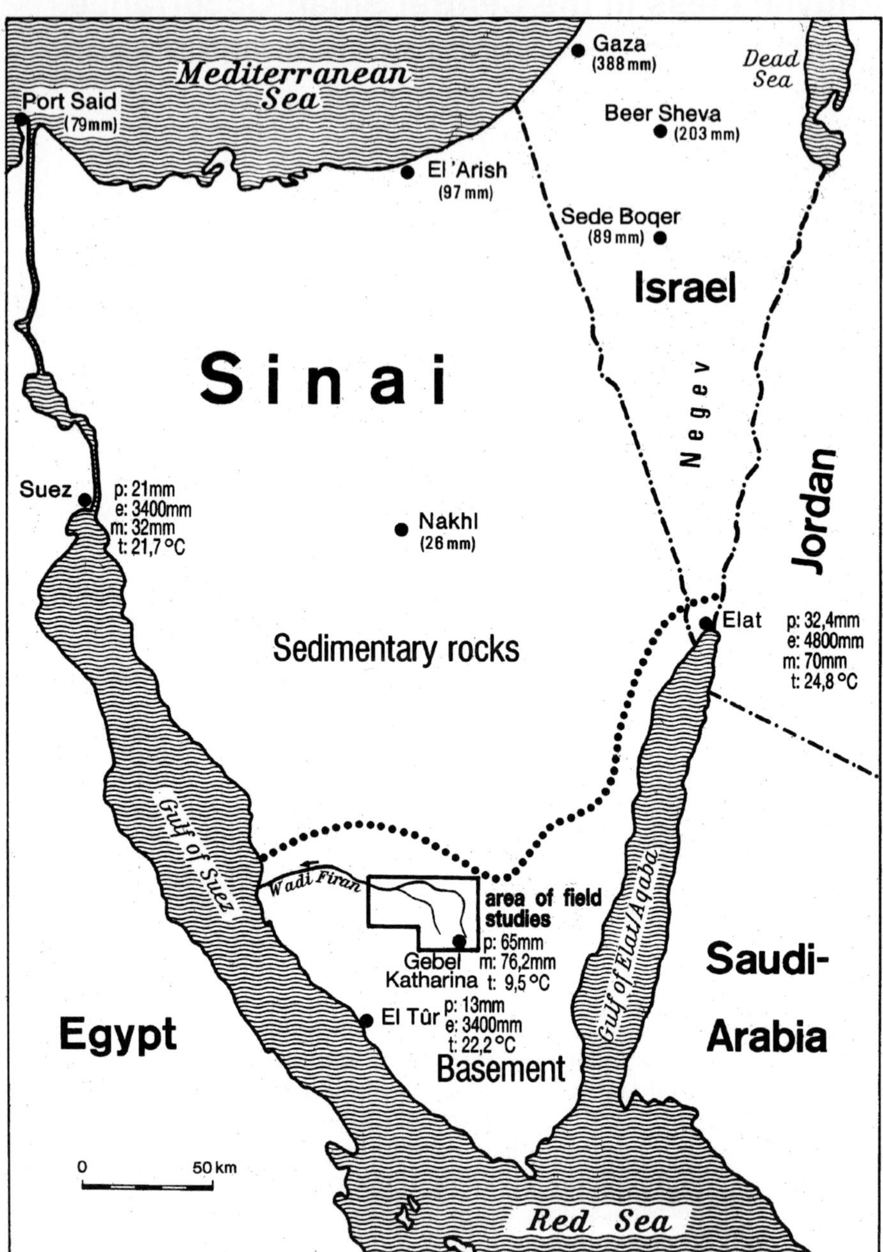

Fig. 1. Location map including the study area, the mean annual precipitation (p) and potential evaporation (e), the maximum rainfall during 24 h (m) in mm and the mean annual temperature (t) in °C

I Introduction

The striking, mighty silt occurrences in the wadis Feiran, Es Sheikh and Solaf in the southern part of the Central Sinai (Fig. 1) have attracted much attention among geoscientists for well over 100 years. Numerous geologists and geomorphologists have studied the outcrops around the oases of Feiran and Tarfat in order to identify the *types* of the layered sediment, and their *origin* and *mechanism* of *sedimentation*. Most of the investigators have interpreted the silts as *lacustrine* sediments (e.g. BARRON 1907, de MARTONNE 1947, AWAD 1951, 1953, ISSAR & ECKSTEIN 1969, NIR 1970, 1974). Only BÜDEL (1954) and KLAER (1962) regarded the silts as fluvial terraces, and the first author studying the silts described them as glacial deposits (moraines, FRAAS 1867).

During the past decade the discussion about the origin and mechanism of sedimentation arose again when – in the course of interdisciplinary studies – geoscientists from Munich (Rögner, geomorphology), Karlsruhe (Smykatz-Kloss, Knabe, mineralogy; Roscher, geology) and Bonn (Zöller, TL dating) published new geochemical, sedimentological and geochronological data (RÖGNER & SMYKATZ-KLOSS 1991a, b, 1993, 1998; EL SHERBINI 1993; SMYKATZ-KLOSS et al. 1997, 1998, 1999/2000, 2000; RÖGNER et al. 1999, KNABE 2000; NAGUIB 2000). The way sedimentation had taken place proved to be especially complex and is still subject to discussion. In the present paper the authors use the new data to assess some possible palaeoclimatological conclusions.

II Field characterization

The sediments occur mainly in the two wadis of Feiran (around the oasis of Feiran) and Es Sheikh (around the oasis of Tarfat) and to a lesser amount in the tributaries of both wadis (Solaf, El Akhdar, Ikhbar and Sayan, see Figs. 2 and 3). They consist of interlayered sequences of (a) coarse-grained and inhomogeneous sands and gravels, clearly fluvial-torrential in origin and therefore not treated in detail in this study, and (b) yellowish silty and quite homogeneous material. The dipping of both types of sediments is different, too: the silts are nearly horizontally layered (1-2° SW), whereas the coarse grained gravels exhibit a stronger inclination. The silts occur at the rims of the wadis, but with some exceptions such as where the sediments are found in the center of the wadi (i.e. at El Bueib and two other locations, see below). The sands and gravels are more likely to occur upstream, near the mouth of tributary wadis and at the bottom positions of former wadi beds.

The reconstruction of the locations of silt deposition results in eleven "basins" (sediment traps). In the case of lacustrine sedimentation, at least ten barriers would have been necessary to create the conditions for lake formation and lacustrine sedimentation (BARRON 1907, de MARTONNE 1947, ISSAR & ECKSTEIN 1969, NIR 1970, RÖGNER & SMYKATZ-KLOSS 1991a, b, 1993). The wadis Es Sheikh/Feiran decline from ~1200 m to sea level, hence the idea of eleven lakes and barriers for the wadi length of ~120 km is hard to imagine.

Fig. 2. Contour map of the study area (modified after the topographical map of Sinai, 1:250000). F = Oasis Feiran, B = El Bueib, T = Oasis Tarfat

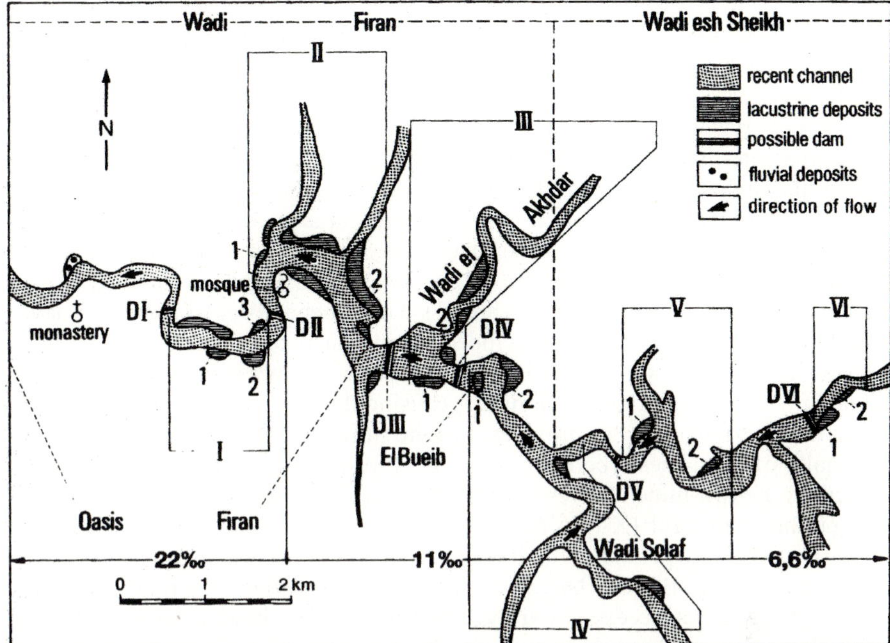

Fig. 3. Occurrence of the fine-grained sediments in the region of the oasis Feiran. I – VI = "basins"; $^0/_{00}$: average gradient of the drainage channel. "Lacustrine deposists" = alluvial loess. Figure from Rögner & Smykatz-Kloss 1991a.

Only in a few occurrences does the existence of a former lake seem to be reasonable, e.g. at the location El Bueib, at the mouth of wadi El-Akhdar (location III) and – probably – at locations I and II at the oasis of Feiran (compare Figs. 2, 3). The other silt occurrences are situated at the rim of the wadis. The thickness of the outcropping silt/gravel profiles is around 20-25 m. Upstream the thickness and the silt/gravel ratios decrease; downstream they increase. Near the small village of the oasis of Feiran the thickness is ~50 m.

III Pleistocene Lakes

Lacustrine environments during the Pleistocene should not generally have prevailed in the central Sinai. Of the few exceptions – where lake-like or swamp conditions are evident – two occur at the rim of the wadi (Feiran: locations II and I in the oasis), a third one (location III) occurs at the entrance of the tributary wadi El Akhdar into the main wadi, and the fourth – El Bueib – could be the only one with a true lacustrine environment for silt deposition (see Fig. 3).

Our location I, the westernmost of all eleven "basins", which includes limnic mud snails in some horizons previously reported upon by Barron (1907), and several travertine beds that evidently precipitated from sewage or muddy waters. The second locality (i.e. "basin" II) is situated at the eastern rim of the small village Feiran, where the northern Wadi Ikhbar joins the main wadi. It shows several indications for a "pseudo-lacustrine" environment. In lower horizons pieces of transparent crystals of gypsum have been found ("Marienglas", "plaster of Paris"), which indicate that formation conditions must have been at least swampy. Some meters above the "Marienglas" horizons two small and dark (black) horizons of 1-3 cm thickness appear. They can be traced over an area of a few hundred square meters. KNABE (2000) found these black layers (II-98-2n and II-98-2p, see Table 1) and a similar third one in "basin" V (i.e. V-97-2, Table 1) enriched in manganese and in some heavy metals (Table 1). The geochemistry of these dark layers points to a former swamp that included an interface of reducing and oxidizing waters (from the tributary and from the main river, respectively). At this interface Eh and pH conditions changed abruptly, causing the precipitation of $Mn^{3+,4+}$ - and – to a lesser extent – Fe^{3+}-hydroxides. In spite of relative high amounts of iron and especially of manganese (between 7 and 15%, respectively, compare Table 1), no crystalline Fe-Mn-minerals were found in these layers. That means the precipitation products are still (X-ray) amorphous, very fine-grained and thus including large surface areas (per weight unit) that have functioned as adsorbers for several heavy metals (Mo, Ba, Co, Ni ..., see Table 1). The milieu of formation of these dark layers was not necessarily truly lacustrine. It may have been that of ponds exhibiting stillwater conditions, probably inactive branches of a former river ("oxbow lakes"). The *source* of the high amounts of Mn probably stems from some nearby outcropping and underlying *amphibolite* dykes, which are quite common in that part of Wadi Feiran (e.g. FRIZ, 1987). Some of the silts at the base of the profiles still contain relict amphiboles. Mn^{2+} is mobilized under reducing conditions (swamps!). The process of accumulation and oxidation of Mn^{2+} (after partial weathering of the amphibolitic dykes) permits a higher residence time for Mn than for iron in water (e.g. 10^4 to 10^3 years, compare HEM, 1964, and WEDEPOHL, 1972). But the kinetics of Mn^{2+} oxidation is strongly influenced by a combination of processes of adsorption and catalysis: apart from autocatalytic deposition of Mn, the Mn precipitation (as $Mn^{4+}O_2$) is catalyzed by surfaces of quartz, feldspars, ferric iron oxides, calcite (WEDEPOHL 1972). Amphibolites contain up to 2000 ppm of Mn (ENGEL et al. 1964).

At *El Bueib* – further up the main river – the silts occur just in the center of the main wadi without any contact to its rims (Fig. 4). The coarse-grained material of fluvial-torrential origin is only deposited at the base of the fine-grained silts (Fig. 4). The well-rounded boulders, gravels and (coarse) sand partly cover a large dyke. It seems that these sediments accumulated as deltaic deposits at the upper end of a lake. In this case there is no necessity to search for a barrier (a dam), because the huge (up to 50 m thick) silt layers of the pseudo-lake II (2 km downstream from El Bueib) could enable the storage of water upstream, at El Bueib. The increasing accumulation of silts at location II forced the backwash of water and led to lacustrine conditions at El Bueib.

Table 1. Chemical enrichment in dark layers (data from KNABE, 2000)

	mass-%		mg / kg								
	Fe_2O_3	MnO	Ba	Sr	As	Co	Ni	Mo	V	U	Pb
V-97-2 location see Fig. 3	8.7	6.14	1,744	1,290	32	140	135	420	270	24	...
Ø of 12 *light* layers (silts)	5.03	0.07	303	370	7	15	19	3	78	6	...
factors of enrichment in V-97-2 vs light layers	1.6	88	6	3.5	4.5	9	7	140	3.5	4	...
II-98 2n	7.05	14.2	7,375	1,288	32	197	256	460	400	18	49
II-98 2p	7.06	9.3	4,500	968	26	180	207	307	300	17	...
(2n + 2p)/2	7.06	11.75	5,940	1,130	29	190	230	385	350	18	49
Ø of 13 light layers	5.14	0.13	250	317	5	13	20	20	65	6	14
factors of enrichment (2n + 2p)/2 vs. light layers	1.4	90	24	3.6	6	15	12	20	5.4	3	3.5

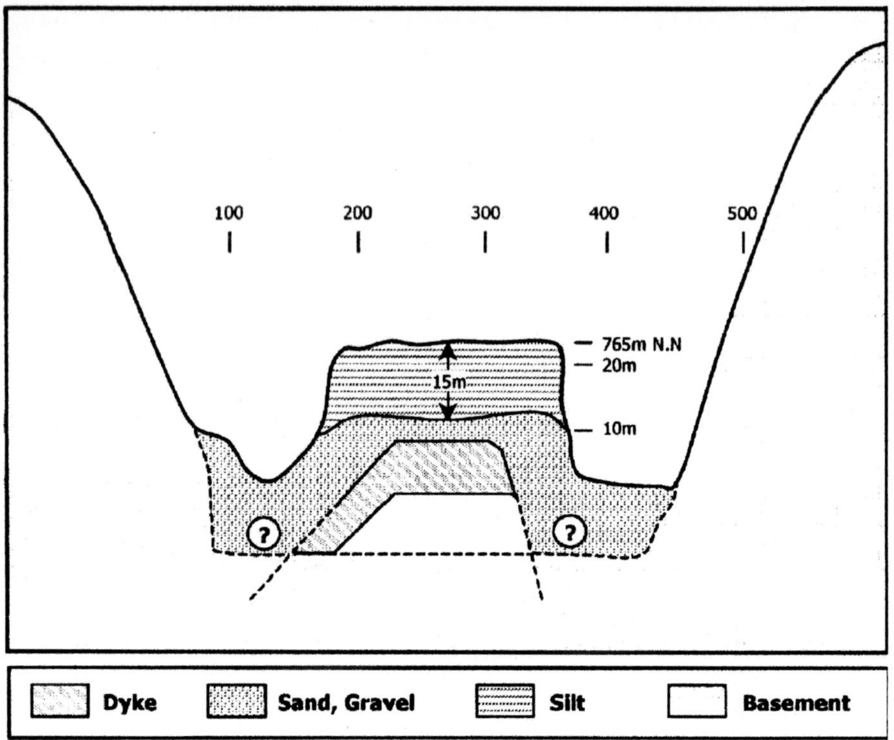

Fig. 4. Cross section at El Bueib

Much further upstream – i.e. at locality XI (east of Tarfat) – an extreme accumulation of calcareous nodules ("Lösskindl") was present, but which is not found anywhere else in the wadi system. In other climatic regions (e.g. more humid) these nodules in loess horizons mirror processes of soil formation, including strong processes of carbonate dissolution and precipitation in lower horizons or groundwater tables. The silty profiles in the wadis of the central Sinai, however, show only the *beginning* of soil formation, indicated by rhizomorphic structures (locations I and II) and low amounts of newly formed iron-hydroxides (goethite). The permeability of the alluvial loess of the Sinai wadis is much lower than that of central European loess occurrences (compare to SMYKATZ-KLOSS et al., this volume). Therefore the striking accumulation of these nodules in location XI (and *only* in XI !) may hint to a period of swampy condition in this basin too. But this seems somewhat questionable due to the local geomorphology (e.g. a very broad basin with larger inclination of the flat ground).

IV Origin of the silts

Before our studies were carried out NIR (1970) was the only author who believed the silts to be aeolian material (loess). He saw the region of origin in the north-eastern direction, i.e. in the escarpment of Tih. But the Mesozoic limestones of the Tih escarpment are compacted and in their mineralogical composition different from the carbonates that occur in the Sinai silts. Especially the crystal chemical composition of the minor carbonate mineral *dolomite* shows some differences: the

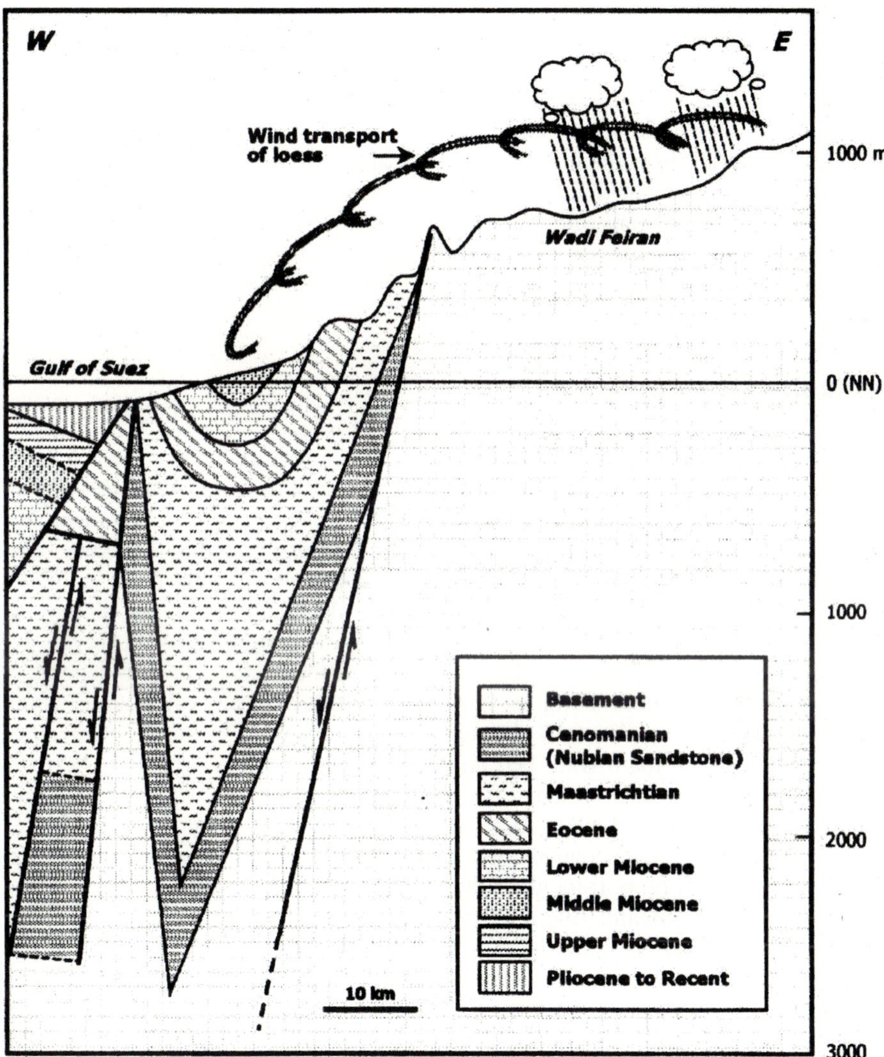

Fig. 5. The Gebel Ataqua structure,situation at the time of loess deposition

rocks of the Tih escarpment contain pure, iron-free dolomite, while the mineral from the Sinai silts includes small, but remarkable amounts of *iron*.

The silts show grain size distributions that are very characteristic for loess (KNABE, 2000). Some horizons of the localities II and V contain remnants of fossils (KNABE, 2000) which B. Reichenbacher and W. Stinnesbeck identified as Miocene foraminifera (personal communication). B. Roscher found similar species in Miocene foraminifera marl of the south-western Arabian Peninsula. From studies of the Red Sea and the Gulf of Suez (BAYER et al. 1988) the sedimentary structures of the Gulf of Suez are well-known (Fig. 5): in the center of the Ataqua structure Miocene globigerina marls occur. This still fairly unconsolidated marl could have been the sediment of origin for the Sinai loess. During the last glacial maximum 20 ka ago the sea-level of the oceans was 100 meters lower than today. This suggests that the Miocene marls occurred at the surface (not being covered by water as before or after the glacial periods), and that the wind could have transported the material up to the hills and wadis of the central Sinai. The aeolian material consists of quartz, low amounts of feldspars, calcite, (Fe-bearing) dolomite, some clay mineral portions *and* foraminifera (in traces). Some layers at the bottom of the silt profiles include traces (up to a few percent) of amphiboles as well, which came into the silts from the underlying ultramafic dykes.

V Mechanism of deposition

With regard to the deposition of the Sinai silts most authors believe them to have formed as *lacustrine* sediments (for a detailed bibliography see RÖGNER & SMYKATZ-KLOSS 1991a, b, 1993 or SMYKATZ-KLOSS et al. 2000). Arguments for this include: the undisturbed and (nearly) horizontal layering, the homogeneity of the grain size, the occurrence of some mud snails (locations I and II), and the occurrence of some varve clays (loc. II). For some of the silt locations this may be true, as has been outlined above. For the majority of the Sinai silts, however, detailed sedimentological investigations showed the mechanism of silt deposition having been much more complex (SMYKATZ-KLOSS et al. 1999/2000; 2000; KNABE 2000). Near localities VI, VII and VIII three wells facilitated the study of the sediments of the main wadi bed down to 10 meters. In two profiles near the rim of the wadi (location V), the silts disappeared after 50 cm. At other localities silts and coarse-grained fluvial sediments showed chaotic, abrupt and discontinuous sequences. The model that seems to explain the sedimentary structures completely is shown in Fig. 6. It includes the different deposits (sediments) of a mixed load river, which is meandering and transporting coarse-grained material (gravel, sand) and in periods of high floods and high transport energies even boulders (rolling on the ground of the river bed), *and* fine-grained (silty) material in suspension. The coarse-grained sediments are the weathering products of rock outcrops in the wadis, upstream from the places where they were finally deposited.

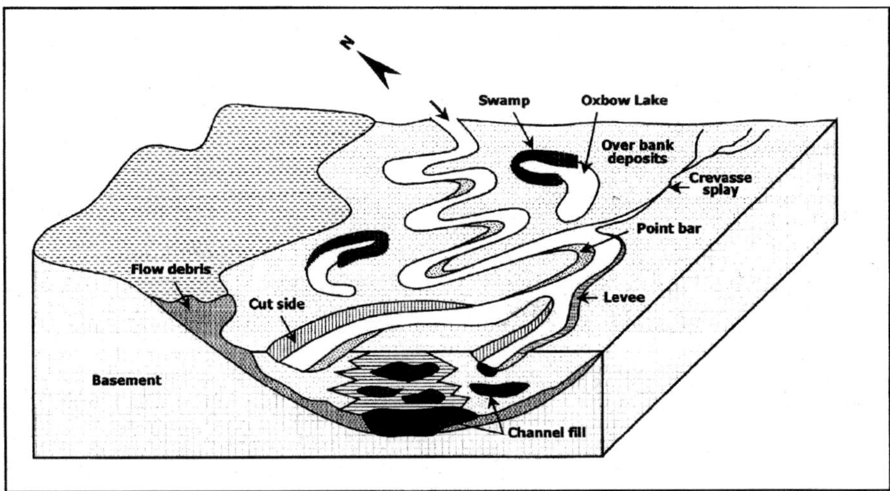

Fig. 6. Sedimentation milieu of a mixed load river

These deposits occur at point bars in river bends, in the former wadi bed when the transport energy had decreased, or at places where the ground of the river is somewhat rough.

As outlined above, the origin of the suspended, silty-clayey material is quite different; it was brought into the wadis from the Gulf of Suez or from even further west by the prevailing western winds. There, a smaller part of it sedimented directly into the wadi bed. The larger portion of this loess sedimented on the slopes of the wadis. Later these silts were washed out by surface overflow and thus brought into the river that transported it westwards, i.e. backwards in the direction to the Gulf. In periods of high floods the fine material passed the banks of the meandering river and sedimented as overbank fines on the plains beside the river. To a lesser extent the silty material came to the sediment plains by crevassing of the lévees (Fig. 6).

After some time the river changed its bed. Now, coarse material was deposited over fine-grained, overbank fines were deposited on point bars and so on, and this occurred many times during the existence of this meandering river. This happened in periods of high water activity, i.e. (semi-) humid environments. During more arid periods water activity decreased and sometimes the river may have dried out (like today, in a hyper-arid environment). In (semi-) humid periods of lower water activity, where no remarkable sedimentation by the (now smaller) river could appear, processes of soil formation could have started, and vegetation appeared around the ox-bow lakes (e.g. in swampy environments).

VI Soil formation on the alluvial loess and palaeoclimatologic considerations

The rhizomorphic structures found in the silts of locations I, II and V are proof for the beginning soil formation and for the abundant vegetation. NAGUIB (2000) observed geochemical as well as mineralogical differences due to chemical weathering during soil formation processes for a profile of nearly 4 m (location II, compare SMYKATZ-KLOSS et al., this volume). The geochemical variations of the silts due to hydrogeochemical processes in the upper parts of the silt layers permit the differentiation between (relatively) dry – yet still humid – and (relatively) humid periods.

The processes of *loess transport* should have occurred in an *arid* (cold) environment. In order to remove the loess from the wadi slopes into the wadis, rain is necessary. Thus a more *humid* period must have occurred some time after the loess was deposited on the slopes. For the processes of sedimentation of overbank fines, high water in the river is needed. But this does not necessarily mean a *humid* period! Today the flooding of allogenous rivers are often observed in (semi-) arid environments, actually in north or south Africa (Nile, Niger, Limpopo). Thus, the sedimentation of overbank fines seemed to occur in semi-arid to semi-humid periods, more or less regularly (annually?) and without extreme amplitudes of precipitation in the regions of sedimentation. Only this period of (overbank) sedimentation can be dated (by means of thermoluminescence technique). The TL dates given below represent the time of (overbank or lacustrine) sedimentation and the following period of soil formation. The soil formation is a process of chemical weathering, a process of interactions between surface and soil waters with the layered silts. The process of soil formation continues as long as soil waters (pore solutions, surface waters) are available. Climatic changes *to more arid conditions halt* soil development. Climatic changes *to more humid* conditions intensify chemical weathering and thus soil development. A change of the flow direction of the river, e.g. by forming another meandering curve, will change the sediment character (from alluvial loess to fluvial sand or gravel).

The studied profiles include many (alluvial) loess horizons, generally not more than 10-20 cm thick. Often – although not generally! – the uppermost 2-4 cm are characterized by clear indications of chemical weathering (= soil formation), compared to the lower part of the same horizon that seems to be unweathered. The *intensity* of chemical weathering reflects the intensity of water activity, assuming that the durations of the weathering periods are comparable. This means, and is the base for the following palaeoclimatological considerations, that high intensity of chemical weathering = a *relatively* humid period and that low intensity of chemical weathering = a *relatively* arid period.

As outlined by SMYKATZ-KLOSS et al. (this volume), the *ratios* between soluble cations (Na^+, K^+, Ca^{2+}, Mg^{2+}, Ba^{2+}, Cu^{2+}, Sr^{2+} etc) and hydrolysate ions (Al^{3+}, Fe^{3+}, Ti^{4+}, V^{5+}), due to their ionic potential, *decrease* when soil waters and fine-grained material in these silts interact. For the alluvial loess of the Sinai, this has been shown by SMYKATZ-KLOSS et al (1998, 1999/2000), KNABE (2000) and NAGUIB (2000). The following relations are compiled from data by KNABE (2000).

Table 2. Chemical analyses of alluvial loess layers and overlying soils, location II in Wadi Feiran; (data from KNABE, 2000) [*l* = alluvial loess; *s* = soil]

Layer	Thickness (cm)	SiO_2 mass-%	Al_2O_3 mass-%	Fe_2O_3 mass-%	Na_2O mass-%	K_2O mass-%	TiO_2 mass-%	Ba mg/kg	V mg/kg
1 s	3	32.06	9.18	5.14	0.51	1.09	0.680	334	85
l*	1	16.72	5.01	7.06	0.28	0.60	0.366	4500	301
2 s	4	35.14	10.13	5.82	0.58	1.28	0.722	208	92
l*	3	18.55	5.41	7.04	0.39	0.64	0.408	7375	401
3 s	1.5	37.24	12.23	6.11	0.53	1.45	0.749	184	79
l	6	31.37	7.09	3.53	0.63	1.02	0.605	358	74
4 s	4	35.83	10.22	5.34	0.54	1.32	0.756	245	80
l	2	41.63	7.69	4.38	0.89	1.21	0.765	304	72
5 s	4	33.02	8.86	4.88	0.54	1.18	0.670	245	77
l	2	42.73	9.43	5.14	0.85	1.41	0.821	293	80
6 s	3	40.27	8.84	5.34	1.08	1.33	0.735	249	79
l	3	43.39	10.10	4.49	1.85	1.39	0.751	398	72
7 s	10	34.51	10.19	5.22	0.59	1.19	0.704	240	82
l	20	54.64	12.78	6.67	2.80	1.82	1.01	569	101
-------	32**								
8 s	4	37.84	8.62	4.97	0.70	1.28	0.717	275	83
l	20	52.07	8.46	5.29	1.21	1.49	0.952	401	91
-------	13**								
9 s	15	39.03	9.58	5.00	0.85	1.39	0.731	266	81
l	4	45.81	8.01	4.49	0.93	1.31	0.800	356	64
10 s	4	36.55	8.65	4.56	0.67	1.20	0.670	254	77
l	4	40.65	8.08	4.30	0.78	1.24	0.627	260	82
11 s	2	32.11	7.95	4.84	0.56	1.03	0.625	241	84
l	5	41.54	7.95	4.72	0.89	1.22	0.743	320	76
-------	15**								
12 s	2	38.40	10.0	5.29	0.64	1.26	0.775	332	82
l	15	48.04	7.79	4.83	1.06	1.27	0.848	360	80
13 s	4	37.27	9.28	4.98	0.71	1.19	0.706	280	73
l	13	50.42	8.50	4.88	1.15	1.42	0.914	393	79

Table 2. (cont.)

14 s	4	36.08	9.88	5.07	0.61	1.10	0.742	226	82
l	9	46.70	7.61	4.34	0.92	1.26	0.796	354	72
15 s	2	34.81	9.50	5.08	0.58	1.16	0.705	254	85
l	10	51.54	10.61	5.34	2.02	1.57	0.919	428	85
16 s	4	37.86	11.0	5.33	0.68	1.35	0.748	235	84
l	8	56.54	7.64	4.50	0.91	1.27	0.803	356	81
17 s	6	40.90	8.37	4.11	0.69	1.16	0.688	273	63
l	60	57.22	7.92	3.64	1.15	1.20	0.607	294	62
(bottom of the profile)									
∅ soils (17)		39.38	9.56	5.12	0.65	1.23	0.713	255	81
∅ alluvial loess (15) (without Mn-rich layers l - 1 and 2)		46.51	8.64	4.70	1.30	1.34	0.797	363	78

* dark layer, extremely rich in Mn, Ba, Sr … (compare table 1)
** interlayers of fluvial material

The *absolute* amounts of these elements and oxides do not show any systematic differences between the substratum (= alluvial loess) and the soil layer that developed from the underlying silt. The amounts of Fe_2O_3, Al_2O_3 and V seem to be enriched in the soils (but not in all!). The amounts of Na_2O, Ba and Cu decreased in the soils. TiO_2 and K_2O behave neutrally, with Ti^{4+}, as a hydrolysate ion, only slightly mobilized. The K^+ has largely been adsorbed on the soil components and has not been removed from the soil such as with Na^+, Ba^{2+}, Cu^{2+}, Mg^{2+} or Ca^{2+}. The earth alkalis (Ca^{2+}, Mg^{2+}) function as soluble cations as well (analogue to Na^+). Yet, for the following geochemical calculations, they will not be considered because the high amounts of Ca, and Mg (Sr) in loess horizons exaggerate the (small) variations between loess and soil (see SMYKATZ-KLOSS et al., this volume).

The systematic variations caused by chemical weathering appear to be much clearer regarding the *ratios* of soluble cation to hydrolysate (Table 3). Horizons 1 and 2 are characterized by very high amounts of Mn, Ba, Sr and V (apart from other heavy metals, see Table 1). Horizon 1 is not chemically weathered at all, and horizon 2 only a little (Table 3). All other horizons of the profile show remarkable indications of chemical weathering, e.g. decreasing ratios of SiO_2/Al_2O_3, SiO_2/Fe_2O_3, Na_2O/TiO_2, Ba/Al_2O_3 and Ba/V. The fact that K^+ is largely retained in the soil by adsorption is mirrored in increasing K_2O/TiO_2 (or K_2O/Al_2O_3) ratios from loess to overlying soil. Relative to Na^+, the K^+ is "enriched" (retained) in the soils too. By comparing the ratios of both types of layers (loess substratum and soil), a "factor of enrichment" is found (Table 3), which is normally <1. That means the processes of soil formation (weathering, leaching) lead to a decrease in the ratios from substratum (loess) to the overlying soil, thus mirroring the (partial) removal of silica, Na^+, Ba^{2+} etc. and the retardation of K^+. The mean values of the seven ratios listed in Table 3 are then taken as a measure for palaeoclimatic reconstructions (Fig. 7, Table 3).

Table 3. Oxide and element ratios from Table 2; factor of enrichment from loess to soil (*f-en s*)

layer	$\frac{SiO_2}{Al_2O_3}$	$\frac{SiO_2}{Fe_2O_3}$	$\frac{Na_2O}{Al_2O_3}$	$\frac{Na_2O}{TiO_2}$	$\frac{Na_2O}{K_2O}$	$\frac{Ba}{Al_2O_3}$	$\frac{Ba}{V}$	$\frac{\Sigma f^7}{7}$	$\frac{K_2O}{TiO_2}$
1 s	3.49	6.24	0.056	0.75	0.47	0....364	3.93	without	1.60
l	3.34	2.37	0.056	0.76	0.47	0...8982	14.95	SiO_2	1.64
f-en s	1.05	2.63	1.00	0.99	1.00	Mn-Ba-V-rich layer		1.01	0.98
2 s	3.47	6.04	0.057	0.80	0.45	-----	-----	without	1.77
l	3.43	2.63	0.072	0.96	0.61			SiO_2	1.57
f-en s	1.01	2.30	0.79	0.83	0.74	Mn-Fe-Ba-rich layer		0.84	1.13
3 s	3.04	6.09	0.043	0.71	0.37	0....150	2.33		1.94
l	4.42	8.89	0.089	1.04	0.62	0....505	4.84		1.69
f-en s	0.69	0.69	0.48	0.68	0.60	0.30	0.48	0.56	1.16
4 s	3.51	6.71	0.053	0.81	0.41	0....240	3.06		1.75
l	5.41	9.50	0.116	1.16	0.74	0....395	4.22		1.58
f-en s	0.65	0.71	0.46	0.70	0.55	0.61	0.73	0.63	1.11
5 s	3.73	6.77	0.061	0.81	0.46	0....276	3.18		1.76
l	4.53	8.31	0.090	1.04	0.60	0....311	3.66		1.72
f-en s	0.82	0.82	0.68	0.78	0.77	0.89	0.98	0.80	1.02
6 s	4.56	7.54	0.122	1.47	0.81	0....282	3.15		1.81
l	4.30	9.66	0.183	2.46	1.33	0....394	5.53		1.85
f-en s	1.06	0.78	0.67	0.60	0.61	0.72	0.57	0.73	0.98
7 s	3.37	6.61	0.059	0.84	0.50	0....236	2.93		1.69
l	4.28	8.19	0.219	2.77	1.54	0....445	5.63		2.55
f-en s	0.79	0.81	0.27	0.30	0.33	0.53	0.52	0.51	0.66
8 s	4.39	7.61	0.081	0.976	0.55	0....319	3.31		1.79
l	6.16	9.84	0.143	1.271	0.81	0....473	4.41		1.57
f-en s	0.71	0.77	0.57	0.77	0.68	0.67	0.75	0.70	1.14
9 s	4.07	7.81	0.089	1.163	0.61	0....278	3.28		1.90
l	5.72	10.20	0.116	1.163	0.71	0....444	5.56		1.64
f-en s	0.71	0.77	0.77	1.0	0.86	0.63	0.59	0.76	1.16

Table 3. (cont.)

10 s	4.23	8.02	0.078	1.00	0.56	0....294	3.3		1.79
l	5.03	9.45	0.097	1.244	0.63	0....317	3.17		1.98
f-en s	0.84	0.85	0.80	0.80	0.89	0.93	1.04	0.88	0.91
11 s	4.04	6.63	0.070	0.896	0.54	...303	2.87		1.65
l	5.23	8.80	0.112	1.198	0.73	...403	4.21		1.64
f-en s	0.77	0.75	0.63	0.75	0.74	0.75	0.68	0.72	1.01
12 s	3.84	7.26	0.064	0.825	0.51	0....332	4.05		1.63
l	6.17	9.95	0.136	1.250	0.83	0....462	4.50		1.50
f-en s	0.62	0.73	0.47	0.66	0.61	0.72	0.90	0.67	1.09
13 s	4.02	7.48	0.077	1.006	0.60	0....302	3.84		1.69
l	5.93	10.33	0.135	1.222	0.81	0....462	4.98		1.51
f-en s	0.68	0.72	0.57	0.82	0.74	0.65	0.77	0.71	1.12
14 s	3.65	7.12	0.062	0.822	0.55	...229	2.72		1.48
l	6.14	10.76	0.121	1.156	0.72	...461	4.88		1.58
f-en s	0.60	0.66	0.51	0.71	0.75	0.50	0.56	0.61	0.94
15 s	3.66	6.85	0.061	0.823	0.50	...267	2.99		1.65
l	4.85	9.63	0.190	2.198	1.29	...403	5.04		1.71
f-en s	0.75	0.71	0.32	0.37	0.39	0.66	0.59	0.54	0.96
16 s	3.44	7.10	0.062	0.909	0.50	...214	2.80		1.81
l	6.09	10.34	0.119	1.133	0.72	...466	4.40		1.58
f-en s	0.57	0.69	0.52	0.80	0.69	0.46	0.64	0.62	1.14
17 s	4.89	9.95	0.082	1.003	0.60	...326	4.33		1.69
l	7.22	15.72	0.145	1.895	0.96	...371	4.74		1.98
f-en s	0.68	0.63	0.57	0.53	0.63	0.88	0.91	0.69	0.86

Table 4. Factors of enrichment from loess to soil (mean values for all 17 studied loess/soil pairs)

	SiO_2	Al_2O_3	Fe_2O_3	Na_2O	K_2O	TiO_2	Ba	V
Ø f-en s	0.85	1.11	1.09	0.50	0.92	0.90	0.70	1.04

The calculation of the mean values for all 17 studied loess/soil pairs is shown in Table 4: only the true hydrolysates Al_2O_3, Fe_2O_3 and V actually increased in amount from loess to soil. K_2O and TiO_2 are relatively immobile, while SiO_2, Ba^{2+} and primarily Na^+ are removed noticeably from the soil horizons.

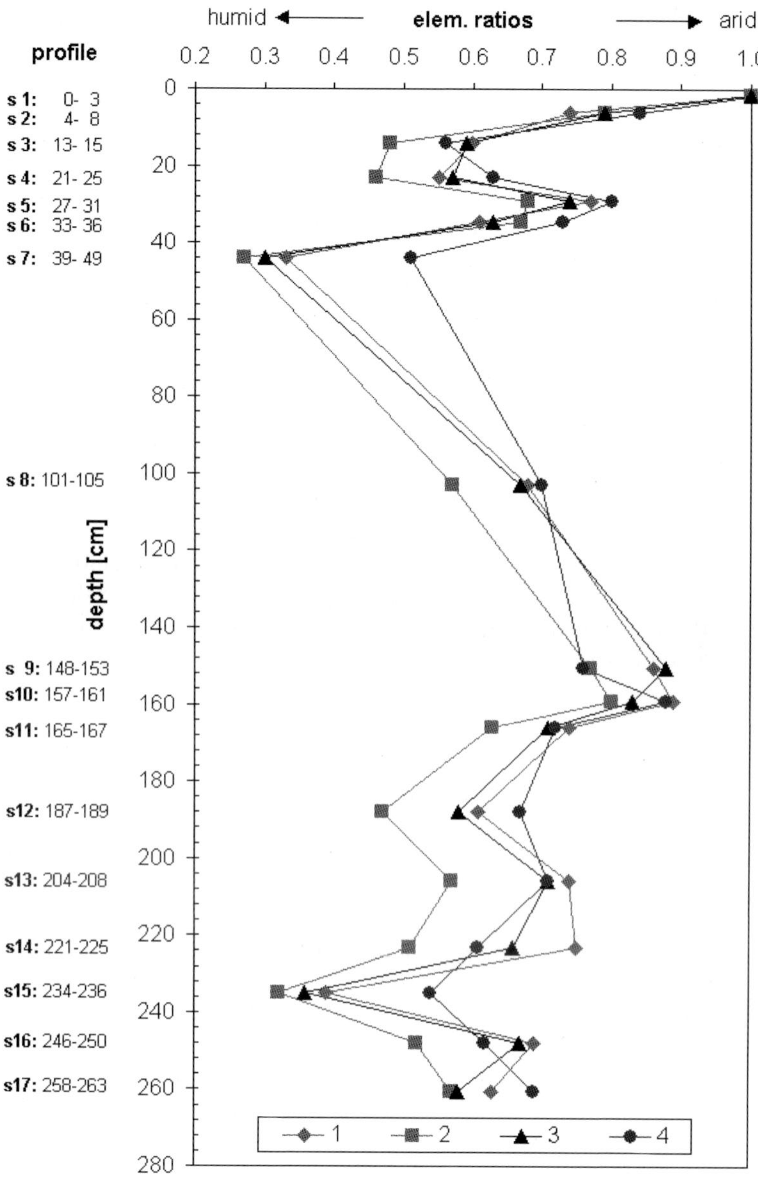

Fig. 7. Variation of oxide ratios in 17 soils of the studied alluvial loess profile (Loc. II, Wadi Feiran);
-1- = Na_2O/K_2O;
-2- = Na_2O/Al_2O_3;
-3- = $(Na_2O/K_2O + Na_2O/Al_2O_3 + Na_2O/TiO_2) / 3$;
-4- = $(Na_2O/K_2O + Na_2O/Al_2O_3 + Na_2O/TiO_2 + SiO_2/Al_2O_3 + SiO_2/Fe_2O_3 + Ba/Al_2O_3 + Ba/V) / 7$

The studied partial profile represents a thickness of 3.20 m. The 17 loess/soil pairs form less than one-tenth of the total outcrop and belong to the lower part of the whole profile. They include indications for several changes in the palaeoclimate. Figure 7 shows soils 7 and 15 to be formed during quite humid periods, while soils 1, 5, 6, 9, 10, 13, 14, and 16 mirror relatively arid environments of formation. The tendency from older to younger horizons goes towards aridity.

Two thermoluminescence (TL) ages were determined from this partial profile, with the samples taken at a vertical distance of 2 m. Both sets of data show an age of 27 ± 4 ka (RÖGNER et al. 1999, SMYKATZ-KLOSS et al. 1999/ 2000).

TL age data are available for four other locations (V, VII, XIa, XIb). The 10 ages (in total) vary between 27 ± 4 and 11 ± 1 ka. These are the ages of silt deposition either as overbank fines or as deposits in swamps, respectively. The data reflect the moment of sediment burial. The thickness of the alluvial loess profiles decreases from west to east, namely from the profiles around the oasis of Feiran (I, II, V) to those east of the oasis of Tarfat (VII, XI). Probably the older silts (> 20 ka) were eroded at the eastern locations, but possibly the older aeolian sediments did not reach the eastern parts of the wadis.

The period between 27 and 11 ka corresponds to marine oxygen isotope stage 2 (= MIS-2). During this period glaciers were at a maximum globally. Consequently, the temperatures were lower globally. The Sinai loess profiles do not include traces of periglacial conditions (permafrost etc.). The climate is assumed to having been colder than today, but not to an extreme. Probably, the climate was that of a steppe.

VII Conclusions

The layered silts of the wadis of the central Sinai are alluvial loesses. They contain Miocene foraminifera, which originate from the center of the Ataqua anticline in the Gulf of Suez, where loose Miocene globigerina marls formed outcrops at the surface 20-27 ka ago. From there, aeolian processes transported the loess eastwards during the global ice-periods and the consequent sea-level decrease until the loess was sedimented on the slopes of the wadis (Feiran, Es Sheikh, Solaf). Later (how much later?), when the climate changed to more humid periods, rain washed out the fine-grained aeolian material and transported it into the main river of the wadi and into the abundant ponds, oxbow-lakes and swamps aside the river. In times of high water of the river the suspended fine-grained, silty material passed over the banks of the meandering river ("overbank-fines") and sedimented on the plains behind the levées. Coarse-grained fluvial-torrential material, e.g. the weathering products of rock outcrops upstream were deposited in front of obstacles on the ground of the river, as point bars in river bends and as alternative layers over the silts when the river had changed its bed. Ten TL data range between 27 and 11 ka, thus indicating that the alluvial loess deposition of the central Sinai took place during the last late glacial.

In the times without sedimentation, which were sufficiently humid to initiate chemical interactions between soil waters and silt material, soil formation occurred. The chemical processes of soil formation caused geochemical changes in the uppermost layer, and rhizomorphic structures in these uppermost layers prove the abundance of vegetation on these soils. The intensity of chemical weathering and thus of soil formation is shown in the geochemical analyses, e.g. in varying ratios of soluble cations to hydrolysates (comparing loess/soil pairs). The mean values of the "enrichment factors from loess to soil" ("f-en s") are reliable as palaeoclimatic indicators. More clearly, the variations of the factors Na_2O/ K_2O or Na_2O/Al_2O_3 (compare Fig. 7) mirror palaeoclimatological changes during the deposition of the alluvial loess: Low ratios point to periods of intensive chemical weathering, that means (relatively) humid periods (e.g. soils 7 and 15); high ratios point to relatively arid periods (e.g. soils 1, 5, 9, and 10, see figure 7). Soil 7 exhibits the greatest thickness of all studied horizons.

Acknowledgements

The authors are grateful to Kirstin Fuhlberg, Natalie Naguib, Frank Friedrich, Georg Istrate (all from Karlsruhe), Abram Bishay (Cairo), Bernhard Eitel (Heidelberg), Klaus Hüser (Bayreuth) and Rajiv Sinha (Kanpur, India) for their company and for many helpful discussions in the field, to Bettina Reichenbacher (Munich) and Wolfgang Stinnesbeck (Karlsruhe) for the identification of microfossils, to Nadine Smykatz-Kloss and Tom Dearnley (Stroud, Gloucestershire) for improving the authors' English, to Beate Oetzel, Maria Tannhäuser and Wolfgang Klinke (Karlsruhe) for drawing the figures and preparing the manuscript, respectively, and – last but not least – to the German Research Foundation for financial support (Ro 84/5, Sm 17/23 and 17/24).

References

Awad, H. (1951): La montagne du Sinai central. Étude morphologique. – Soc. Roy. Geogr. Égypte, Le Caire, 247 p.

Awad, H. (1953): Signification morphologique des dépôts lacustres de la Montagne du Sinai central. – Bull. Soc. Roy. Geogr. Égypte XXV, 23-28

Barron, T. (1907): The topography and geology of the Peninsula of Sinai (Western Portion). – Survey Dept., Cairo, 241 p.

Bayer, H.-J.; Hötzl, H.; Jade, B.; Roscher, B. & Voggenreiter, W. (1988): Sedimentary and structural evolution of the northwest Arabian Red Sea margin. – Tectonophysics 153, 137-151

Büdel, J. (1954): Sinai, "die Wüste der Gesetzesbildung" als Beispiel für die allgemeine klimatische Wüsten-Morphologie. – Abh. der Akad. f. Raumforschung und Landeskunde (Mortensen-Festschrift)

El-Sherbini, M. (1992): Composition and palaeoclimate of the Pleistocene lacustrine sediments of Wadi Feiran, Sinai. – Proc. 3rd Conf. Geol. Sinai f. Development, Ismailia 1992, 153-160

Engel, A.E.J.; Engel, C.D. & Havens, R.G. (1964): Mineralogy of amphibolite interlayers in the gneiss complex, NW Adiron Mts., NY. – J. Geol. 72, 131

Fraas, O. (1867): Aus dem Orient – Geologische Beobachtungen am Nil, auf der Sinai-Halbinsel und in Syrien. – 222 S., 3 Taf., Verlag Ebner & Seubert, Stuttgart

Friz, A. (1987): Mineralogie und Geochemie pan-afrikanischer Ganggesteine der südlichen Sinai-Halbinsel. – Diss. Fak. Bio-Geowiss., Univ. Karlsruhe

Hem, J.D. (1964): Chemistry of manganese in natural waters. – U.S. Geol. Survey Water Supply Papers 1667-B

Issar, A. & Eckstein, Y. (1969): The lacustrine beds of Wadi Feiran, Sinai: Their origin and significance. – Isr. J. Earth Sci. 18, 21-28

Klaer, W. (1962): Untersuchungen zur klimagenetischen Geomorphologie in den Hochgebirgen Vorderasiens. –Heidelberger Geogr. Arbeiten 11

Knabe, K. (2000): Sedimentpetrographische und geochemische Untersuchungen an Sedimenten im Wadi Feiran (Südsinai) zur Klärung ihrer Herkunft und Ablagerungsbedingungen – Paläoklimatische Überlegungen. – Diss. Fak. f. Bio- und Geowissenschaften, Univ. Karlsruhe

Martonne, E. de (1947): Reconnaissance géographique du Sinai. – Ann. Géogr. LVI, 241-264

Naguib, N. (2000): Mineralogische und geochemische Untersuchungen an einem Schwemmlöss-Profil in der Oase Feiran, Sinai (Ägypten). – Dipl.-Arbeit, Fak. f. Bio-u. Geowissenschaften, Univ. Karlsruhe

Nir, D. (1970): Les lacs quaternaires dans la region de Feiran (Sinai Central). – Rev. de Géographie Physique et de Géologie Dynamique (2), vol. XII, 335-346

Nir, D. (1974): Lacustrine / fluviatile sediments in Feiran and Tarfat el Kudrein. – Z. f. Geomorphologie N.F. Suppl. Bd. 22, 32-34

Rögner, K. & Smykatz-Kloss, W. (1991 a): The deposition of aeolian sediments in lacustrine and fluvial environments of Central Sinai (Egypt). – Catena Suppl. 20, 75-91

Rögner, K. & Smykatz-Kloss, W. (1991 b): Fluviale Geomorphodynamik im Zentralen Sinai während des jüngeren Quartärs. – Freiburger Geograph. Hefte 33, 209-221

Rögner, K. & Smykatz-Kloss, W. (1993): The fine-grained sediments of Wadi Feiran (Sinai, Egypt): Origin and sedimentology. – Z. f. Geomorphologie N.F., Suppl.-Bd. 88, 123-139

Rögner, K. & Smykatz-Kloss, W. (1998): The fine-grained loess-like sediments of the Wadi Feiran, Sinai, Egypt: Possibilities of palaeoclimatic interpretations? – In: Alsharhan, A.S., Glennie, K.W., Whittle, G.L. & Kendall, C.G.St.C. (Eds.): Quaternary Deserts and Climatic Change. Balkema, Rotterdam, pp. 209-211

Rögner, K.; Smykatz-Kloss, W. & Zöller, L. (1999): Oberpleistozäne paläoklimatische Veränderungen im Zentral-Sinai (Ägypten). – Erdkunde Bd. 33, 220-230

Smykatz-Kloss, W.; Knabe, K. & Rögner, K. (1997): The geochemical development of (semi-) aridic soils as a tool for the reconstruction of palaeoclimatic changes: a case study for the Wadi Feiran, Sinai, Egypt. – Zbl. Geol. Paläontol. I, 41-57

Smykatz-Kloss, W.; Knabe, K.; Rögner, K.; Hüttl, C. & Zöller, L. (1998): Palaeoclimatic changes in central Sinai. – Palaeoecology of Africa 25, 143-155

Smykatz-Kloss, W.; Roscher, B.; Knabe, K.; Rögner, K. & Zöller, L. (1999/2000): Wüstenforschung und Paläoklimatologie im zentralen Sinai. – Chemie d. Erde 59, 245-258

Smykatz-Kloss, W.; Roscher, B. & Rögner, K. (2000): Gab es im Sinai pleistozäne Seen? – Regensburger Geographische Schriften 33 (Heine-Festband), 127-139

Smykatz-Kloss, W.; Smykatz-Kloss, B.; Naguib, N. & Zöller, L. (this volume, in print): The reconstruction of palaeoclimatological changes from mineralogical and geochemical compositions of loess and alluvial loess profiles. – In: Smykatz-Kloss, W. & Felix-Henningsen, P. (Eds.): Palaeoecology of Quaternary Drylands

Wedepohl, K.H. (1978): Manganese. Part B-O in: Wedepohl, K.-H. (Ed.): Handbook of Geochemistry II-3

The reconstruction of palaeoclimatological changes from mineralogical and geochemical compositions of loess and alluvial loess profiles

Smykatz-Kloss, Werner[1)], Smykatz-Kloss, Bettina[2)], Naguib, Natalie[3)] & Zöller, Ludwig[2)]

1) Institute for Mineralogy and Geochemistry, University of Karlsruhe, D-76128 Karlsruhe, Germany

2) Geographical Institute, University of Bonn, D-53115 Bonn, Germany

3) Research Center Karlsruhe for Technology and Environmental Sciences, D-76021 Karlsruhe, Germany

ABSTRACT

The study is concerned with the reconstruction of palaeoclimatological changes in (alluvial) loess profiles by means of mineralogical and geochemical criteria. For this, the intensity of chemical weathering (palaeosoil formation) is taken from the ratios of soluble cations to hydrolysates (e.g. Na_2O/Al_2O_3; K_2O/TiO_2) or the degree of corrosion of minerals. Examples of study are (a) a part-profile of alluvial loess in the Wadi Feiran, Sinai, and (b) five loess profiles from the region around Bonn, Germany. Changes in palaeo-humidity are mirrored by geochemical variations, changes in palaeotemperatures by the amount and types of clay minerals.

INTRODUCTION

Chemical weathering of minerals and rocks means *hydrolysis*. As much as processes and factors of physical weathering (variations in temperature, mechanical stress) may split, fragmentize or break down a primary rock - such as for example granite,- without the interaction with surface water or pore water no chemical changes take place (Correns 1968, Berner 1971).

This equally means if in a young but otherwise mineralogical and petrographical homogenous sediment profile zones are found which indicate chemical changes, these must be zones of higher water activity.

If diagenetic mineral re-formations (which also cause chemical changes) can be excluded from the analysed profiles, then (and this is the basis for further arguments) these observed zones of higher water activity must be indicators of more humid phases after deposition took place. The basis (i.e. no diagenetic mineral rock-re-formations) applies to young, soft sediments, e.g. to many loesses.

"Loess" itself is already a palaeo-climatic indicator, its aeolic transport having taken place in a (cold-) arid environment, such as during times of Pleistocene glaciation. The types of alluvial loess of the Sinai analysed here (see below) also fall into this time period according to thermo-luminescence dating (27-11 ka, see Rögner et al. 1999, Smykatz-Kloss et al. 1999/2000). However, the periods of loess formation were frequently interrupted by volcanic activity (sedimentation / deposition of tephra and tuffs) or by times where no aeolic sediment transport took place.

These 'resting periods' can be caused by climate, and in more humid periods processes of chemical weathering, i.e. starting soil formation, caused by interactions between (rain-) water and loess should take place. It should be possible to trace both, the interstratification of (possibly already reformed) tuffs as well as the palaeo-soil formation by means of mineralogical, particularly geochemical, criteria. This will be attempted in the following, with the example of the (re-deposited) types of loess of the southern Sinai-peninsula, which since their deposition up to today have been predominantly exposed to (hyper)-arid climate (compare Stanley 1994, Knabe 2000). As a second example types of loess of the Bonn region will be examined, which after their deposition have been exposed to predominantly more humid climates. The related geochemical and mineralogical data are taken from research carried out by Natalie Naguib (alluvial types of loess of the Sinai) and Bettina Smykatz-Kloss (Pleiser Hügelland, Bonn region), which should be referred to for further details.

PRELIMINARY REMARK ON PHYSICS

According to their grain size the analysed types of loess are silts (~0.063 mm in diameter). They are well distributed and fairly homogeneous in their mineral composition of quartz, plagioclase, potassium feldspar, mica, calcite and dolomite. In the loamy horizons carbonates are lacking. This process of (horizons) becoming loamy, in which the carbonates calcite and dolomite have been dissolved, has not affected the silicate-based loess components as the solubility of quartz and silicates is much lower than that of carbonates. Moreover, the permeability of silts is quite low. Permeability values (Kf) have been determined in the Geological Institute Karlsruhe for a typical alluvial loess from the Sinai (9 measurements of different sample areas at temperatures of 19-21°C). For these temperatures the Kf-values lie between 1.1 x 10^{-8} and 1.5 x 10^{-8} m/s. As the temperatures during deposition of the Sinai loesses will have almost certainly been below 20°C (glaciation !), the Kf-values will have been even much lower.

The silt therefore nearly acts as a barrier. In more humid regions much bigger pathways for descending solutions may be caused by greater water activities and the resulting faster dissolving of carbonate. The at times of sedimentation quite substantial matrix-carbonate- contents, which are still present in many loess profiles (of the Pleiser Hügelland), however, indicate that even a great part of these loesses of a mainly humid region (after loess deposition) did not experience any substantial water activities (other than on the surface). Yet, exceptionally large loesskindl enrichments can be found in the hyper-arid Sinai loesses at a few locations (compare Rögner et al., in this book), which indicate strong processes of carbonate- re-solution.

At these few locations in the Wadi Es Sheikh special geomorphological conditions must have prevailed (compare Rögner et al. in this book). But even here palaeo-climatic deductions can be construed from the geochemical results - as some preliminary analyses have shown (compare Rögner& Smykatz-Kloss 1991a, b; Smykatz-Kloss et al. 1999/2000).

PRELIMINARY REMARK ON CHEMISTRY

According to their behaviour in watery solutions, ions can be classified into 3 categories: (I) soluble cations, (II) hydrolysates and (III) soluble anion-complexes. The mode of solubility is controlled by the *ionic-potential* (= ion charge / ion radius): Cations with an ion potential of <3 belong to the soluble ions (all alkalis, earth alkalis, Fe^{2+}); during chemical weathering of prime-minerals the alkalis etc which are resolute to being soluble are carried away in the solution. With increasing radius, however, the mobility of the ions decreases and the tendency to a high degree of adsorption to fine-grained sediment or soil particles increases. In this way large cations (K^+, Rb^+...) are increasingly taken out of watery solutions through adsorption, whereas the Na^+ (which is small and protected additionally by means of hydrate sheaths) and the -although bivalent- yet even smaller Mg^{2+} get carried away - resolute to being soluble - into the sea. Potentials above 10 mark soluble anion - complexes (SO_4^{2-}, CO_3^{2-}, BO_3^{3-}...). Ions with a potential between 3 and 10 precipitate quickly from watery solutions as hydroxides (hydrolysates: Al^{3+}, Fe^{3+}, Ti^{4+}). Although the mode of solubility of single ions is influenced by further parameters (presence of particular anions, activities of solubility companions, pH value, Eh, temperature), fairly reliable patterns of behaviour emerge for the 'simple' ions (alkalis, earth alkalis, Al^{3+}, Ti^{4+}) (Berner 1971, Baes and Mesmer 1976, Lindsay 1979, Drever 1982, Brookins 1988, Krauskopf &Bird 1995, Smykatz-Kloss et al. 1998).

As criteria for a palaeo-soil within a loess (-loam)- profile therefore *lower ratios* of a soluble cation (Na, K, Rb, Ca, Mg., Fe^{2+}, Sr, Ba) to a relatively insoluble hydrolysate (Al_2O_3, Fe_2O_3, TiO_2) can be applied- lower ratios in comparison to those layers that lie less weathered or lie immediately above. However, not all possible ratios are equally suitable: the observed ratios *should only have been changed through chemical processes operating during soil formation* and not through diagenetic processes or processes of increasing loaming. This means that for samples containing carbonate Ca, Mg and Sr can only be used in a limited way as chemical changes caused by weathering of primary feldspars and micas are masked by the very high contents of Ca, Mg and Sr of the carbonate minerals. In carbonate-free samples, however, these elements can be taken into consideration. Only in feldspars and micas (and not in carbonates) are Na^+, K^+, Rb^+, Al_2O_3, TiO_2, Fe^{2+}, Fe^{3+}, Mn^{2+} situated, so that changes in the ratios of these cations (or oxides) are true criteria for processes of hydrolysis and hence for chemical weathering and soil formation. Here again, certain constraints have to be made for iron and manganese only, as their behaviour is complex (influence of Eh, bacterial reduction or oxidation) and they get built partially into soil and sediment carbonates (Fe-dolomite, ankerite, kutnahorite, compare Böttcher 1993, Brannath, 1995). Yet (palaeo-) soil horizons often show increased Fe^{3+} contents.

The loss of alkalis and earth alkalis caused by chemical weathering (and therefore the *decrease* in ratios of soluble cations / hydrolysates) is hence - as mentioned above - the more noticeable the smaller the soluble cation, i.e. Na, Mg, Fe^{2+}>Ca>K, Rb>Sr, Ba. Gallet et al (1998), too, used the ratios of Na_2O/Al_2O_3 and K_2O/Al_2O_3 of loesses as a "chemical index of alteration". First attempts to use loess geochemistry for palaeo-climatic interpretations of loess-profiles, have been carried out mainly at the monumental Chinese loesses (Liu, 1985, 1991; Derbyshire et al. 1995; Liu et al. 1993; Gallet et al. 1996).

The ratio of the two hydrolysates TiO_2 and Al_2O_3 proves to be almost constant in the sediment profiles (for example in the loesses examined here) (Gallet et al. 1998, Smykatz-Kloss et al. 1999/2000). It is thus uniform throughout the entire profile II of the Feiran alluvial loesses 0.095 ± 0.015, even with variable intensities of weathering (Smykatz-Kloss et al. 1999/2000, 2000; Knabe 2000). In the loess- profiles of the Pleiser Hügelland it only varies from its constant values in the (Eltville) tuff horizons, in which case it changes to higher values (see below). Palaeo-soils are thus particularly clearly characterised by (lower) values in Na_2O/Al_2O_3 or Na_2O/TiO_2, less clearly by lower values in K_2O/Al_2O_3 and K_2O/TiO_2 as well as (in carbonate-free horizons) by lower MgO/Al_2O_3 and MgO/TiO_2 values. Due to the partial adsorption of the K^+ to clay minerals and organic substances the K^+ is not removed as strongly from the partly weathered horizons as the Na^+, so that the resulting increase in K_2O/Na_2O ratios is characteristic for chemical weathering as well.

PALAEO-CLIMATIC RECONSTRUCTION FROM THE ALLUVIAL LOESSES OF THE SINAI

The thick stratified yellow silts which occur in the wadis of the southern Sinai are not lacustrine formations (except for few locations, see Rögner et al., this book), as it has been assumed by various authors (literature see Rögner & Smykatz-Kloss 1991 a, b, 1998; Smykatz-Kloss et al. 2000), but *alluvial loesses*, deposits of a meandering river on its flood plains ("overbank fines", less often "crevasse splays"). Through changes in the river bed they occur in a continuous alternate stratification with coarser fluviatile sands (compare Smykatz-Kloss et al. 1999/2000, 2000; Knabe 2000). According to their *grain size distribution and their distribution coefficient* these are typical loesses (Knabe 2000) even partly in the silt layers, where they are mixed with fluviatile fine sands. During their deposition as alluvial loesses 15-25 ka ago (Rögner et al. 1999, Smykatz-Kloss et al. 1999/2000, Rögner et al., this volume) the climate was more humid than today (today it is hyper-arid). Since the Pleistocene it has not changed *continuously* but has changed several times between (semi-) arid and (semi-) humid instead (Smykatz-Kloss et al. 1998, 1999/2000, Rögner et al., this volume). The up to 50 metres thick alluvial loess profiles of Wadi Feiran and its side valleys represent a tremendous palaeo-climatic archive for the last 20-30 ka (Rögner et al. 1999), of which only a small part has been analysed so far. Consequently, Knabe (2000) and Naguib (2000) have been able to work on a part-profile of a total of 8m. The results of Naguib (2000) will be discussed *as an example* below.

The 3.5 m thick part-profile analysed by N. Naguib (2000) from the loess-"basin" II at the east edge of the Feiran oasis (compare Rögner & Smykatz-Kloss 1991 a, b; Knabe 2000) connects *directly above* to the approximately 5m thick base-profile (Knabe 2000, Rögner et al., this volume). It consists of 17 alternate sequences of coarse and fine carbonate-containing sands and silts, which are thicker at the base and becoming thinner towards the top. Naguib (2000) divides the 3.5 m into 5 zones, the samples of which differ clearly in their mineralogical and granulometric criteria (Table 1).

The degree of roundness of the quartzes, the degree of preservation of the minerals (determined in thin sections of soil) and the content of authigenous Fe-minerals correlate with trace geochemical data (Figures 1-5). This results in a palaeo-climatic interpretation (see below).

Figures 1-5 show the *mean* values of some geochemical criteria for zones V-I.

According to the mineralogical (Table 1) and geochemical characterisation (Figures 1-5) zones V and I are similar, as are zones IV and II. V and I show the smallest chemical and mineralogical changes with the middle zone III showing the largest. II and IV represent transitional zones between those areas of the profile that have hardly been changed chemically (I, V) and those that have been changed strongly (weathered III). Thus ratios K_2O/Al_2O_3- (Figure 1) and SiO_2/Zr (Figure 2) decrease from V and I to III by a third. SiO_2/TiO_2 (Figure 3) and Ba/Zr (Figure 4) show the same picture. Opposing to this are the contents in *organic carbon*

Table 1. Mineralogical characterisation of the loess part-profile (3.5 m) from the eastern edge of the Feiran oasis (Sinai)

Zone	Degree of roundness of quartz	Degree of preservation of minerals	Fe-(hydr-) Oxide content
I (top)	subangular to subrounded	biotite, feldspars and hornblende rel. fresh	low
II	subrounded to rounded	biotite partly corroded feldspars partly sericitized	noticeable
III	rounded	biotites strongly altered feldspars fully sericitized hornblende absent	large
IV	subangular to subrounded	few biotites partly sericitized hornblende little opacitized	distinguishable
V (bottom)	subangular to subrounded	micas, feldspars and hornblende without corrosion	low

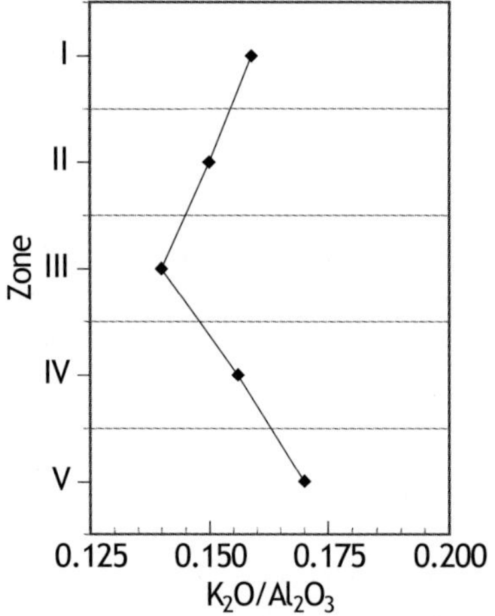

Fig. 1. K_2O/Al_2O_3 ratios (mean values) of the zones I (top) – V (bottom)

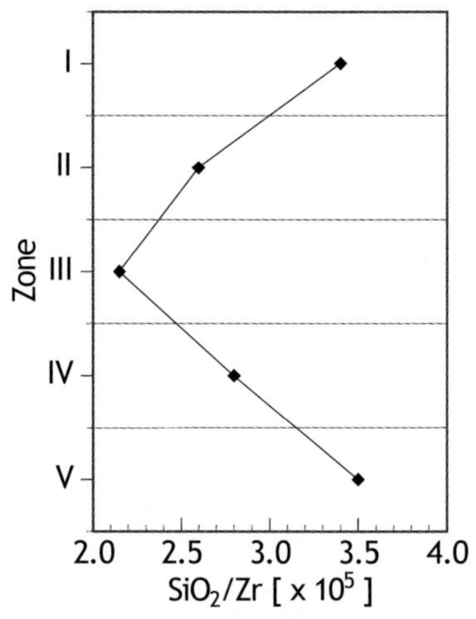

Fig. 2. SiO_2/Zr ratios (mean values) of the zones I (top) – V (bottom)

(Figure 5): they are lowest in I and V and drastically highest in III. Obviously changes in the named ratios, in organic carbon content and in correlative transformations in mineralogy and physical crystallography reflect (palaeo-) climatic milieu changes, which happened in a depositional time span of a few thousand

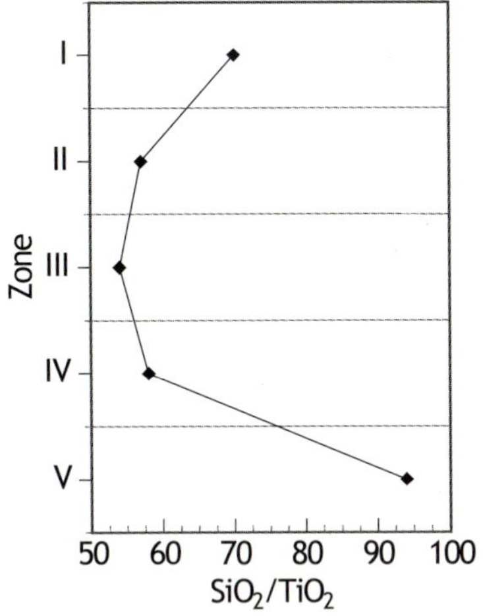

Fig. 3. SiO_2/TiO_2 ratios (mean values) of the zones I (top) – V (bottom)

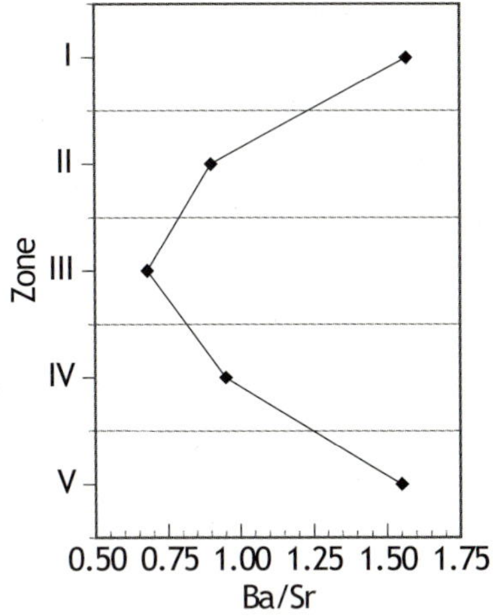

Fig. 4. Ba/Sr ratios (mean values) of the zones I (top) – V (bottom)

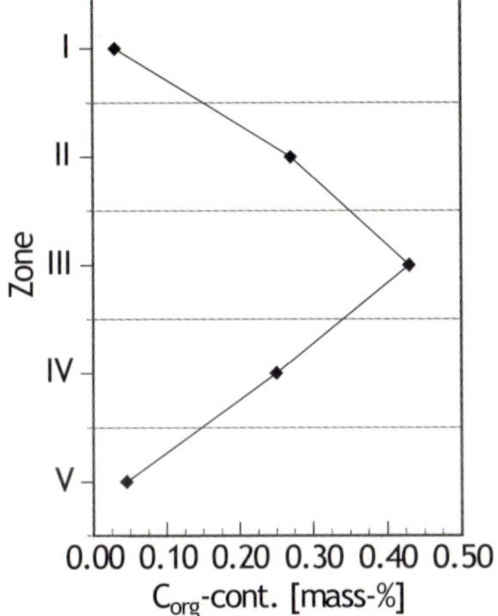

Fig. 5. C_{org} contents (mean values) of the zones I (top) – V (bottom)

years: starting from the slightly changed base zone V water activity (precipitation) and hence intensity of chemical weathering increase continuously, until they reach a maximum in III and decrease again continuously towards I. The analysed 3.5m thick part-profile (~1/14th of the total, 18 ka encompassing 50 meter profile) reflects first an increase in humidity (V → III) and eventually a continuous decrease in humidity towards I. Zone III represents a true palaeo-soil.

PALAEO- CLIMATIC RECONSTRUCTION FROM LOESS OF THE PLEISER HÜGELLAND (BONN REGION)

Contrary to the alluvial loesses of the central Sinai, which experienced nearly continuously an arid milieu, the deposits of the Pleiser Hügelland (and those of

Neustadt/Wied) represent typical central European loesses which experienced humid climates from their deposition onwards. This can already be detected in their mineral composition, in which feldspars and mica have already been transformed to a good degree into stratified silicates (smectite, illite, mixed layers, seldom into kaolinite). Through the deposition of tuffs (such as the Eltville-tuff) new feldspars were added (especially sanidine, but also plagioclases), and the predominantly humid post-sedimentary milieu intensified soil formation and the concentration of organic material.

By looking solely at mineralogical criteria the profile *"Neustadt/Wied"* can be divided into nine zones (B. Smykatz-Kloss, in prep.). Changes in the types of feldspar are striking: Only in the tephra zone (II) and in the deeper humus zone (VI) do potassium feldspars dominate, otherwise plagioclase is the determining feldspar. The actual loess is limited to roughly 4 metres (IV). Humus zone (VI) is fairly rich in clay minerals. In both humus zone (VI) as well as in the *lying* zone (VII) kaolinite occurs, whereas all other zones are free of kaolinite but contain smectite nearly throughout.

Humus zone (VI) shows the lowest contents of Na_2O/SiO_2, MgO/Al_2O_3, MgO/TiO_2, CaO/Al_2O_3, CaO/TiO_2, MgO/Fe_2O_3, Sr/Ba, Sr/Rb and Cu/As and resembles the upper recent soil (I) in mineral composition and chemism. The tephra zone (II) is characterised by low ratios of Na_2O/Al_2O_3, Na_2O/TiO_2, K_2O/Al_2O_3, K_2O/TiO_2, CaO/MgO and very high ratios of K_2O/Na_2O.

According to some of these characteristics quantities of tuff can also be detected in the loess (IV). The decrease of alkalis, earth alkalis and strontium in the recent soil (I) in comparison to the loamy substrate zone (III) would certainly be much more distinguishable, if it had not been superimposed by the tephra weathering (Fe_2O_3, K_2O, MgO, CaO, P_2O_5, TiO_2). Particularly distinct are the (relatively speaking) extremely high P_2O_5 and MnO contents in the tephra zone, the MnO again showing parallels to the humus zone (Figs. 6 and 7).

Samples could be taken from the profile *"Rauschendorf"* of the upper 6 metres. According to its chemism it can be divided into 4 horizons, which can be detected particularly clearly at the oxide ratios, the Sr/Ba and the Sr/Rb ratios. The loss of Na, Mg, Ca and Sr as well as the concentration of Fe characterises zone I as a *soil.*

The decalcified zone II above the loess (IIIb) represents a horizon that has become very loamy. In comparison to the "pure" non-weathered loess (IIIb) IIIa (1.35 m), too, is partly weathered and shows a starting soil formation. Zone IV has to be characterised as an oxidation horizon of a *gley* according to its high Fe concentration and its transformations (decreases) of some easily soluble components (alkalis, earth alkalis, strontium). The TiO_2/Al_2O_3 ratio proves to be consistently constant (0.07).

The profile *"Birlinghoven"* has an outcrop of a depth of 18 metres. The whole profile contains carbonate and comprises Eltville tuff (IV). Plagioclase is the more common feldspar, except for the tuff horizon and the region directly above it, where the potassium feldspar content is greater than the plagioclase content due to additional volcanogenic sanidine. Next to mica, chlorite and smectite small kaolinite contents are also to be observed.

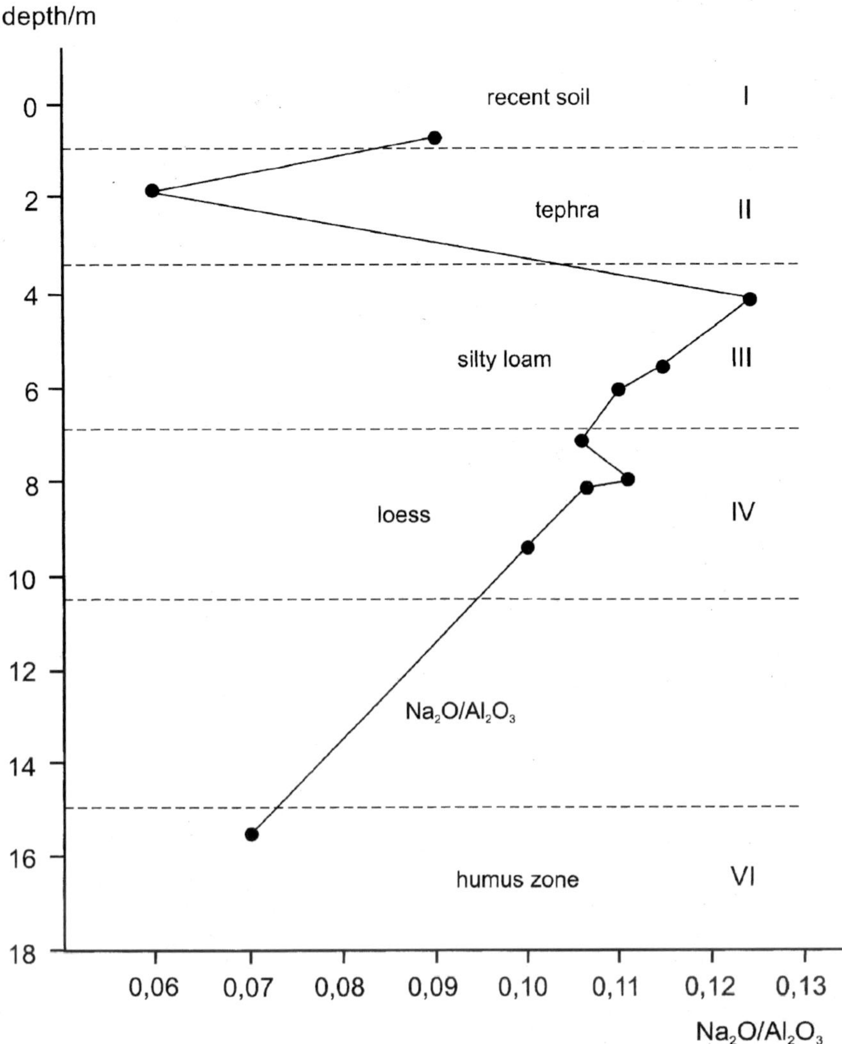

Fig. 6. Na_2O/Al_2O_3 versus depth, profile Neustadt/Wied

This Eltville tuff is the reason for the variable TiO_2/Al_2O_3 values. The tuff is particularly enriched with SiO_2, TiO_2, Fe_2O_3, Ba and Zr. Oxide and element ratios allow a differentiation of the profile into 5 categories. For the upper (recent) soil horizon a noticeable enrichment in K is characteristic (Figs. 8, 9 and 10).

At the "*Thomasberg*" a fairly complex profile had been exposed over 32 metres. The loess has become very loamy and is mixed with numerous tuffs. This is the mineralogical conclusion from the often varying character of feldspars: high tuff components cause a dominance of potassium feldspar (sanidine). The (ex-)

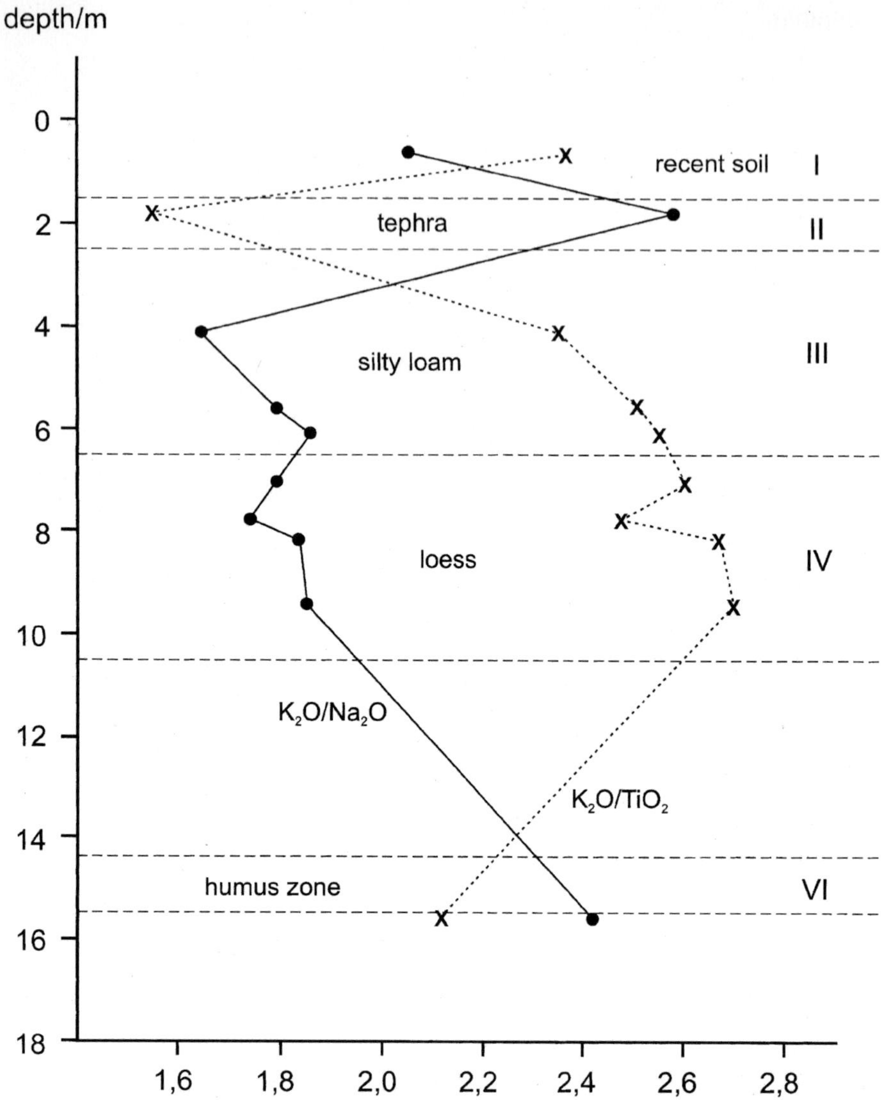

Fig. 7. K_2O/Na_2O (•) and K_2O/TiO_2 (×) vs depth, profile Neustadt/Wied

pyroxenes of the trachyte tuff and a part of the feldspars have been transformed to stratified silicates, the main one of the latter being a three-layered clay mineral with the capability to expand. It expands from 14.5 Å (water saturated) to 17.0 Å (with ethylene glycol). Because of this it is, according to Brindley & Brown (1980), a smectite (montmorillonite). In the upper soil zone (I) the montmorillonite has already "weathered" into an irregular illite/smectite-mixed layer. The

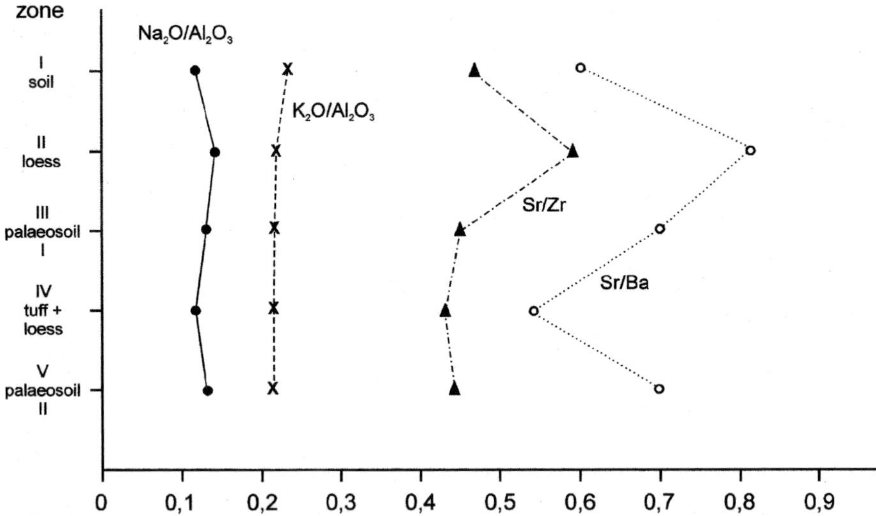

Fig. 8. Mean values for the different zones, profile Birlinghoven, for Na_2O/Al_2O_3 (●), K_2O/Al_2O_3 (×), Sr/Zr (▲) and Sr/Ba (**O**)

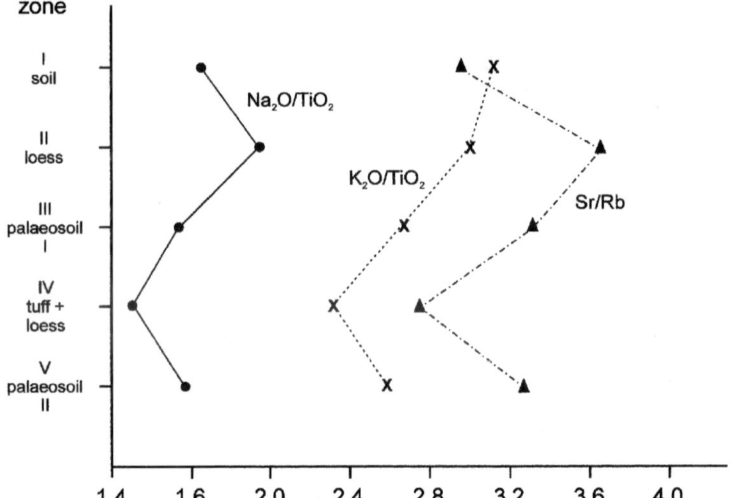

Fig. 9. Mean values for the different zones, profile Birlinghoven, for Na_2O/TiO_2 (●), K_2O/TiO_2 (×), Sr/Br (▲)

complexity of both, petrogenetic as well as weathering events is reflected in the profile differentiation: above a nearly quartz-free but potassium feldspar-rich trachyte tuff (19.5-32 m) lies an approximately 5 metres thick zone of weathered trachyte tuff (IV, very rich in potassium feldspar and smectite), and above that there

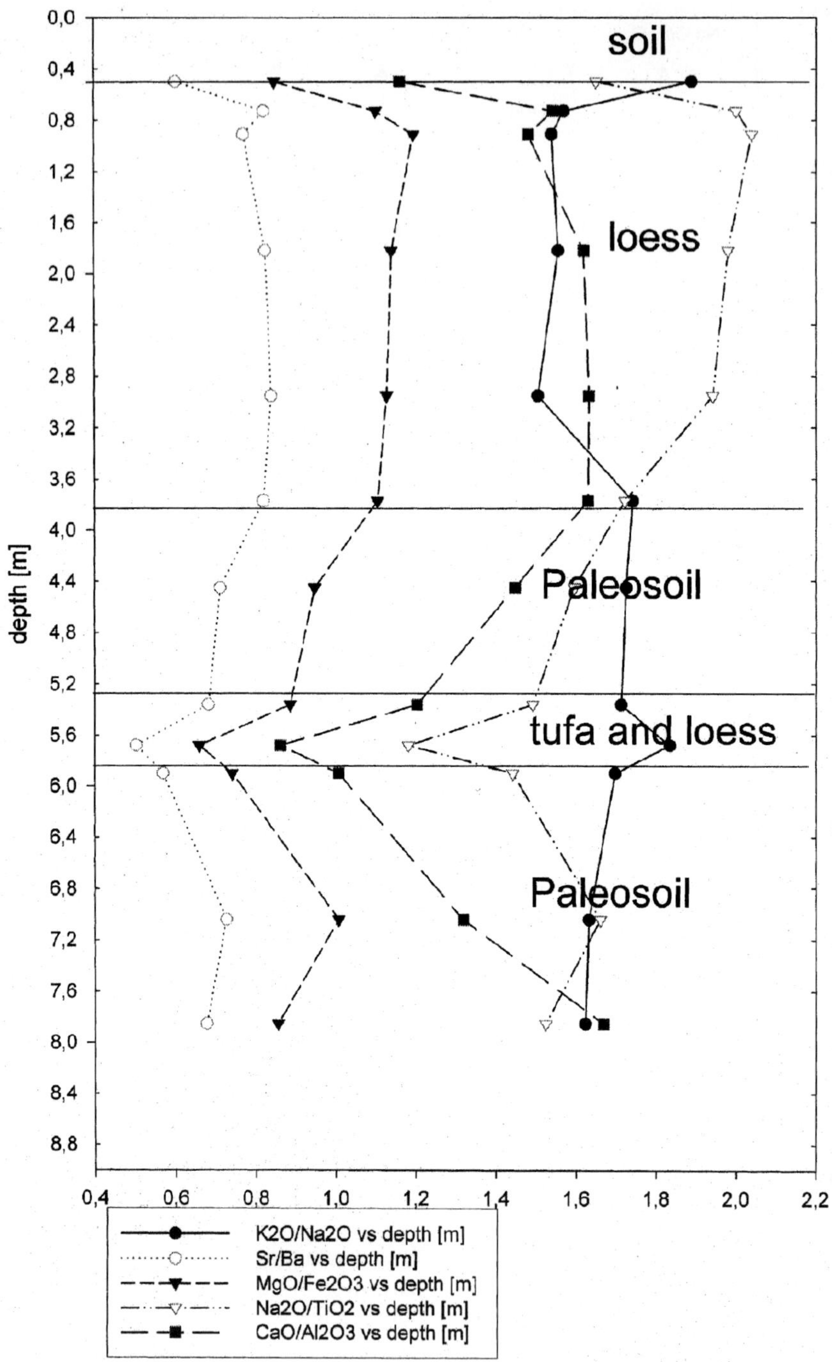

Fig. 10. Chemical ratios, profile Birlinghoven

is calcite- rich loess (III, 9.2 to 12m). Above that is an approximately 7 metres thick zone of loess-loam with varying contents of tuff (II, 1.6 to 9.2m). Depending on the tuff-loam ratio either potassium feldspar (sanidine, tuff) or plagioclase (loam) is dominant. Chemical analyses of zones I-IV reflect their complex petrography. Compared to the other loess-profiles (Neustadt/Wied, Rauschendorf, Birlinghoven, Bockeroth) the systematic loss of alkalis and earth alkalis caused by weathering from the substrate to the soil that originated from it seem to "fail". The usually very constant TiO_2/Al_2O_3 ratio varies considerably (between 0.06 and 0.10). Only after having considered the varying ratios of potassium feldspar / plagioclase caused by the *addition* of volcanogenic potassium feldspar (and the subsequent variations in Na, K and Al_2O_3) do reliable criteria for the recent soil formation become visible (CaO/TiO_2, MgO/TiO_2). The (weathered) trachyte-tuff (IV) itself is characterised by the highest contents of Na_2O, MnO, Rb, Zr and Ba as well as the lowest ratios of TiO_2/Al_2O_3, Fe_2O_3, K_2O/Al_2O_3, MgO/Al_2O_3.

The *drilling Bockeroth* has a depth of 20 metres. Above a loess layer (V) 3 to 4 metres loam appear (IV), of which the top first metre (IVa) already shows clear signs of soil formation. Above this are another 3 to 4 metres loess (III) and recent soil (I).

Zone IV is striking by extremely low contents of MgO and CaO but fairly high contents of SiO_2, TiO_2 and K_2O; it can be assumed that it contains noticeable quantities of an acid tuff (for details see B. Smykatz-Kloss, in prep.).

DISCUSSION

The geochemical profiles of the analysed loess-deposits of the Bonn region (Neustadt / Wied, Rauschendorf, Birlinghoven, Thomasberg, drilling Bockeroth) as well as the alluvial loess deposits of the Feiran oasis (Sinai), allow a differentiation into zones of different water activity, which is expressed as chemical transformations of particular loess horizons. Thus the analysed part-profile of the Feiran oasis reflects initially an increase in chemical weathering and hence in humidity (V → III) followed by a steady decrease to the youngest, fully-arid zone I (compare Figures 1-5, Table 1).

The over 40 chemical analyses (main and trace elements) of the Bonn loess profiles show five petrographical units: loess, loam, lime soil, soil and tuff embedments. Immediately above the Eltville tuff (see profile Birlinghoven) the (clay and loam) layers are still somewhat mineralogically (contents of potassium feldspar) and geochemically "contaminated" by tuff layers (higher ratios of K_2O/Na_2O, TiO_2/Al_2O_3, Fe_2O_3/Al_2O_3, but *lower* ratios of Na_2O/TiO_2, Na_2O/Al_2O_3, K_2O/TiO_2, MgO/Al_2O_3, CaO/Al_2O_3, Na_2O/SiO_2, MgO/Fe_2O_3, MgO/TiO_2, CaO/TiO_2, Sr/Ba, Sr/Rb and CaO/MgO). They show higher contents of potassium feldspar and - where relatively strongly weathered - clear contents of smectite. The smectite of these tuffs is generally a *montmorillonite*. Some horizons include tuff contents, which can only be detected by higher ratios of TiO_2/Al_2O_3, Fe_2O_3/TiO_2,

Sr/Zr, Sr/Rb or Cu/As (such as for example for trachyte tuffs) or by enrichments in CaO, MgO for (carbonate-free!) basaltic tuffs.

However, no palaeo-climatic information can be gained merely by the presence of tuffs. This is only then possible when chemical weathering is recognised in particular depths (and in recent soil), and, as results obtained so far show, in carbonate-rich layers as well (for example in the profile Rauschendorf or in the profile Birlinghoven). Thus in the profiles Neustadt/Wied and Bockeroth one palaeosoil can be demonstrated next to the recent soil whereas in profiles Rauschendorf and Birlinghoven *two* fossil soils can be demonstrated respectively for each profile. The 41 chemical analyses show a fairly constant picture for the mean values of oxide or element content of the 5 units of the loess profiles: Thus the ratio of the two hydrolysates TiO_2 and Al_2O_3 proves to be very constant throughout all profiles (0.076 - 0.077) (exception: tuff and the really complex profile of Thomasberg), so that both can be used as 'reference contents' for the remaining components. In general soil weathered chemically shows (in relation to its substrate: loam for carbonate-free soils, loess for soils containing carbonate, i.e. lime soils) *lower* ratios in all analysed oxide or element pairs - except K_2O/Na_2O and CaO/MgO ratios (and the latter ratio only in lime-free horizons). These two exceptions can be explained either by the size of ions or by chemical reactivity: both, K^+ and Ca^{2+} are bigger than Na^+ and Mg^{2+} respectively, so that they (in comparison to Na^+ or Mg^{2+}) get adsorbed stronger to fine clay mineral or humus substances in the soil and in this way accumulate in the soil in comparison to the substrate. The presence of soil water containing slight amounts of sulfate or HCO_3^- would retain the Ca partially in the soil through precipitation.

According to this the decrease of Fe (II), Na, Mg, Ca, Sr, (K) in the soil (compared to the substrate) relatively to the hydrolysates Al and Ti is caused by chemical weathering (hydrolysis) and marks a period of greater water activity, i.e. a more humid climate phase. The presence of fossil soils is thus an indication for humid phases.

Estimating the size of such *palaeo-humidity* periods is not that simple, because duration of the humid soil formation periods as well as temperature, too, have an influence on the intensity of chemical weathering. This "duration" might possibly be sketched by thermo-luminescense dating (at least roughly), as Zöller did for the Sinai. This resulted in time spans of approximately 18 ka (10-28 ka old) for the whole (alluvial) loess deposition (Rögner et al. 1999, Smykatz-Kloss et al. 1999/2000, 2000; Knabe 2000; Rögner et al. this volume). The dating of the palaeo-soils of Bockeroth, Birlinghoven, Rauschendorf and Neustadt/Wied could lead to the parallelisation of sedimentological events in this loess region around Bonn, further mineralogical analyses for the determination in origin of the loesses and to possible changes in the intensity of loess transports and their respective periods.

For the rough estimation of *palaeo-temperatures* the *clay minerals* have to be examined more closely, which shall be demonstrated at the example of the Neustadt/Wied profile. From the nine different zones four are relatively poor in stratified silicates (chlorite, mica, smectite and kaolinite) namely loam zones VIII and V, loess zone IV and tephra-zone (II).

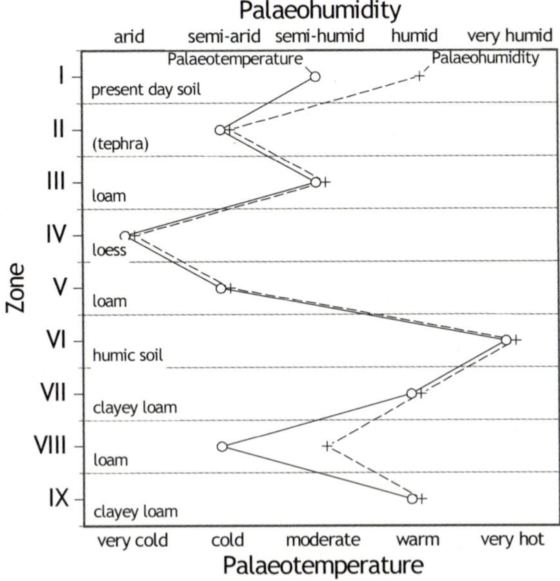

Fig. 11. Reconstruction of the palaeoclimate, Neustadt/Wied

The remaining zones, contrary to the named ones which are poor in clay minerals, are all rich in smectite. The zone richest in clay minerals, "humus zone" VI, is the only horizon that contains additionally to the stratified silicates chlorite, smectite and mica , which occur in all zones, some kaolinite (as the clays directly above and below the "humus zone"). Kaolinite is the typical clay-mineral of warm, humid soils of the tropics. Smectites (montmorillonite, saponite, etc), however, are not only formed in soils and sediments, i.e. by weathering, but also by autohydrothermal re-formation of young *vulcanites*, namely even in those regions without chemical weathering, such as in the basalts of the antarctic peninsula (Blümel et al. 1985). To distinguish between the smectites formed hydrothermally and those formed by chemical weathering in soils or sediments (kaolinite, mica, ...) the *degree of structural disorder* of the stratified silicates is estimated (either by X-ray, thermoanalysis, infrared spectroscopy or electronmicroscopy) (Smykatz-Kloss & Althaus 1975): stratified silicates formed hydrothermally, i.e. at relatively high temperatures at around 80-150°C are generally better structured. According to this criterium the smectites of the Pleiser Hügelland (namely those of all analysed profiles) are true products of chemical weathering, and so are the mixed-layers and kaolin minerals, - except the well-ordered smectite of the Thomasberg profile, which shows to be formed by hydrothermal transformation of volcanogenic material (tuff, see B. Smykatz-Kloss, in prep.). This means: the content of smectite is a measurement for the intensity of chemical weathering and thus for *palaeo-temperatures* when chemical weathering and soil formation took place. According to this zones IX, VII and particularly humus zone VI, to a lesser extent also the sub-recent zone I were relatively warm periods.

Regarding the water activity at the time of formation of clay minerals (by weathering), it is lowest for the formation of smectites and highest for two-layer alumosilicates (kaolinites). The other clay minerals (illite, illite/smectite mixed layers, chlorites partly) exhibit water activities being in-between (Correns 1968).

Thus, the *palaeo-humidity* (or aridity) may be estimated from the total content of newly formed stratified minerals others than smectites, it rises with an increased content of authigeneous (= lesser structured) stratified silicates. This means that the nine zones of the 33 m profile of Neustadt/Wied reflect six changes in the palaeo-climate (compare Figure 11).

Rögner et al. (this volume) demonstrate the palaeoclimatic changes for a larger part-profile of the alluvial Sinai loess, obtained by the outlined geochemical criteria.

Acknowledgements

The authors are very grateful to Dr. U. Kramar for XRF-analyses, to Dr. K. Roehl (Geol. Inst. Karlsruhe) for the determination of the K_f-values, to Maria Tannhäuser, Beate Oetzel and Wolfgang Klinke for preparing the manuscript (all from the Institute for Mineralogy and Geochemistry, University of Karlsruhe), to Nadine Smykatz-Kloss (Stroud, Gloucestershire) for correcting the English, and to the German Research Foundation for financial support (DFG: Sm 17/23, Sm 17/24, Zo 51/15).

References

Baes, C.F. & Mesmer, R.E. (1976): The Hydrolysis of Cations.- Wiley & Sons, New York

Berner, R.A. (1971): Principles of Chemical Sedimentology.- McGraw Hill, New York

Blümel, W.D.; Emmermann, R. & Smykatz-Kloss, W. (1985): Vorkommen und Entstehung von tri-oktaedrischen Smektiten in den Basalten und Böden der König-Georg-Insel (S-Shetlands / West-Antarktis).- Polarforschung 55 (1), 33-48

Böttcher, M.E. (1993): Die experimentelle Untersuchung Lagerstätten – relevanter Metall-Anreicherungsreaktionen aus wässrigen Lösungen unter besonderer Berücksichtigung der Bildung von Rhodochrosit ($MnCO_3$).- Diss., Univ. Göttingen

Brannath, A. (1995): Mineralogisch-geochemische Untersuchungen an Carbonatmineralen und Quarzen aus Eisen-Manganvorkommen in Hessen und Rheingrabenrand-Sulfidvorkommen in Baden.- Diss., Univ. Karlsruhe

Brindley, G.W. & Brown, G. (1980): Crystal Structure of Clay Minerals and their X-ray Identification.- Mineralogical Soc., 5, London

Brookins, D.G. (1988): Eh-pH Diagrams for Geochemistry.- Springer-Verlag, Berlin – Heidelberg – New York

Correns, C.W. (1968): Einführung in die Mineralogie. 2.Aufl., 458 S., Springer-Verlag, Heidelberg

Derbyshire, E.; Kemp, R. & Meng, X. (1995): Variations in loess and paleosols properties as indicators of paleoclimatic gradienst across the Loess Plateau of North China. Quatern. Sci. Rev. 14, 681-697

Drever, J.I. (1982): The Geochemistry of Natural Waters. Prentice-Hall, New York

Gallet, S.; Jahn, B.M. & Torii, M. (1996): Geochemical characterisation of loess-paleosol sequence from the Luochuan section, China, and its paleoclimatic implications.- Chem. Geol. 133, 67-88

Gallet, S.; Jahn, B.M.; van Vliet-Lanoë, B.; Dia, A. & Rosello, E. (1998): Loess geochemistry and its implications for particle origin and composition of the upper continental crust. EPSL 156, 157-172

Knabe, K. (2000): Sedimentpetrographische und geochemische Untersuchungen an Sedimenten im Wadi Feiran (Südsinai) zur Klärung ihrer Herkunft und Ablagerungsbedingungen – Paläoklimatische Überledungen. – Diss. Fak. f. Bio- und Geowissenschaften, Univ. Karlsruhe

Krauskopf, K.B. & Bird, D.K. (1995): Introduction to Geochemistry. 3rd Ed., McGraw-Hill, New York

Lindsay, W.L. (1979): Chemical Equilibria in Soils.- Wiley & Sons, New York

Liu, C.Q.; Masuda, A.; Okada, A.; Yabuki, S.; Zhang, J. & Fan, Z.L. (1993): A geochemical study of loess and desert sand in Northern China: Implications for continental crust weathering and composition.- Chem. Geol. 106, 359-374

Liu, T.S. (1985): Loess and the Environment. China Ocean Press, 251 p., Peking

Liu, T.S. (1991): Loess, Environment and Global Change, Beijing Sci. Press, 288 p.

Naguib, N. (2000): Minerlogische und geochemische Untersuchungen an einem Schwemmlöss-Profil in der Oase Feiran, Sinai (Ägypten). – Dipl.-Arbeit, Fak. f. Bio- u. Geowissenschaften, Univ. Karlsruhe, unveröff.

Rögner, K.; Knabe, K.; Roscher, B.; Smykatz-Kloss, W. & Zöller, L. (2003): Alluvial loess in the Central Sinai: Occurrence, origin, and palaeoclimatological consideration.- In: Smykatz-Kloss, W. & Felix-Henningsen, P. (Eds.): Palaeoecology of Quaternary Drylands, p.81-101

Rögner, K. & Smykatz-Kloss, W. (1991 a): The deposition of eolian sediments in lacustrine and fluvial environments of Central Sinai (Egypt). – Catena Suppl. 20, 75-91

Rögner, K. & Smykatz-Kloss, W. (1991 b): Fluviale Geomorphodynamik im Zentralen Sinai während des jüngeren Quartärs. – Freiburger Geograph. Hefte 33, 209-221

Rögner, K. & Smykatz-Kloss, W. (1998): The fine-grained loess-like sediments of the Wadi Feiran, Sinai, Egypt: Possibilities of palaeoclimatic interpretations? – In: Alsharhan, A.S., Glennie, K.W., Whittle, G.L. & Kendall, C.G.St.C. (Eds.): Quarternary Deserts and Climatic Change. Balkaema, Rotterdam, pp. 209-211

Rögner, K.; Smykatz-Kloss, W. & Zöller, L. (1999): Oberpleistozäne paläoklimatische Veränderungen im Zentral-Sinai (Ägypten). – Erdkunde Bd. 33, 220-230

Smykatz-Kloss, B. (2003): Diss. in preparation, Fak. Naturwiss., Univ. Bonn

Smykatz-Kloss, W. & Althaus, E. (1975): Experimental investigation of the temperature dependence of the "crystallinity" of illites and glauconites.- Bull. Groupe franç. Argiles XXVI, 319-325

Smykatz-Kloss, W.; Knabe, K.; Rögner, K.; Hüttl, C. & Zöller, L. (1998): Palaeoclimatic changes in central Sinai. – Palaeoecology of Africa 25, 143-155

Smykatz-Kloss, W.; Roscher, B.; Knabe, K.; Rögner, K. & Zöller, L. (1999/2000): Wüstenforschung und Paläoklimatologie im zentralen Sinai. – Chemie d. Erde 59, 245-258

Smykatz-Kloss, W.; Roscher, B. & Rögner, K. (2000): Gab es im Sinai pleistozäne Seen? – Regensburger Geographische Schriften 33 (Heine-Festband), 127-139

Stanley, S.M. (1994): Historische Geologie. Eine Einführung in die Geschichte der Erde und des Lebens.- Spektrum, Heidelberg.

Geochemical implications for changing dust supply by the Indian Monsoon system to the Arabian Sea during the last glacial cycle

Dirk C. Leuschner[a], Frank Sirocko[b], Georg Schettler[a], and Dieter Garbe-Schönberg[c]

[a]GeoForschungsZentrum Potsdam, Projektbereich 3.3, Telegrafenberg, 14473 Potsdam, Germany

present address: Institut für Geophysik und Geologie, Universität Leipzig, Talstrasse 35, 04103 Leipzig, Germany.
e-Mail: dcleu@rz.uni-leipzig.de

[b]Institut für Geowissenschaften, Johannes Gutenberg Universität Mainz, Becherweg 21, 55099 Mainz, Germany

[c]Geologisch-Paläontologisches Institut, Christian-Albrechts-Universität Kiel, Olsenhauerstr. 40-60, 24118 Kiel, Germany

Abstract

Element concentrations of 43 elements as well as inorganic and organic carbon content of sediment core 70KL from the western Arabian Sea were measured with high (1 cm) sample resolution. Principal components of the sediment's chemical composition were determined with the help of statistical principle component analysis. These components are representing the major environmental factors at the site. The most important processes controlling the observed variations are the changing lithogenic influx derived from the major wind systems of the region (i. e., the Arabian northwesterly winds, the northeast winter monsoon and the southwest summer monsoon), summer monsoon associated upwelling and biogenic productivity as well as the redox conditions at the sediment-water interface.

The variations of these components show quasiperiodic oscillations in the Dansgaard/Oeschger band (1000-3000 ka) with a significant presence during the entire last glacial. The dominating periods are near 2000 and 1200 years for the variations in the summer monsoon activity, 950 year variations in the winter monsoon and near 1450 and 1050 year oscillations in the dust content transported by the Arabian northwesterly winds.

Introduction

The chemical composition of Arabian Sea surface sediments shows geochemical provinces due to processes of aeolian transport, river discharge, authigenic formation of Mn- and Fe-nodules, hydrothermal activity and preservation of organic matter under anoxic conditions (Shankar *et al.*, 1987; Sirocko, 1995; Sirocko *et al.*, 2000). The geochemical distribution pattern of four time slices, namely the surface sediment, the early Holocene, the Termination 1a and the Last Glacial Maximum are given in Sirocko et al. (2000). Based on this mapping the evaluation of core 74KL, spanning the last 20.000 years, has shown high frequency oscillations at 1785- (upwelling), 1450- (Arabian dust flux), 1150- (upwelling) and 950-years (winter monsoon) (Sirocko *et al.*, 1993; Sirocko *et al.*, 1996). It also demonstrated the presence of teleconnections between high northern latitudes, where a period of 1470 years is the dominant frequency in the oxygen isotope record of the Greenland GISP2 icecore (Grootes and Stuiver, 1997), and tropical/subtropical regions. Such millennial scale (Dansgaard-Oeschger scale) climate oscillations at low latitude sites were first documented in the eastern Pacific Santa Barbara Basin (Behl and Kennett, 1996) and are also well preserved in Arabian Sea sediments (Reichart *et al.*, 1998; Schulz *et al.*, 1998; Prins, 1999; Leuschner and Sirocko, 2000).

In this study high resolution geochemical records from core 70KL are used to evaluate relative variations of dust transport by the Indian Monsoon System to the Arabian Sea during the entire last glacial cycle.

Sedimentation and sediment composition at this site are controlled mainly by the seasonally changing Indian Monsoon system. During the summer warm and moist air from the Central Indian Ocean high pressure cell is attracted by the Tibetan and Central Asian heat low. In the Arabian Sea region this particular process results in a permanent low level jet-like wind (Findlater jet) crossing the Arabian Sea from eastern Kenya to India (Findlater, 1969; McGregor and Nieuwolt, 1998). Associated with these summer monsoon winds are dust plumes reaching from East Africa into the western Arabian Sea, and strong open ocean and coastal upwelling beneath the jet and along the Oman coast, respectively. The summer monsoon is the main contributor of moisture to the Indian subcontinent and therefore responsible for the large majority of river discharge to the Arabian Sea. During boreal winter cool air sinks down over southern Asia and is forced towards the southern Indian Ocean by the pressure gradient forming the northeast monsoon over the Arabian Sea. During this time aeolian dust transport from India to the northwest Arabian Sea is visible in satellite images (Sirocko and Sarnthein, 1989). Nevertheless, the main proportion of aeolian dust to the Arabian Sea is transported by the Arabian northwesterly winds during the summer. These winds last from March/April to October/November with the maximum of dust load being carried in July. They start as low level winds over the Arabian deserts where they entrain large amounts of dust and rise up to midtropospheric heights (up to 5000

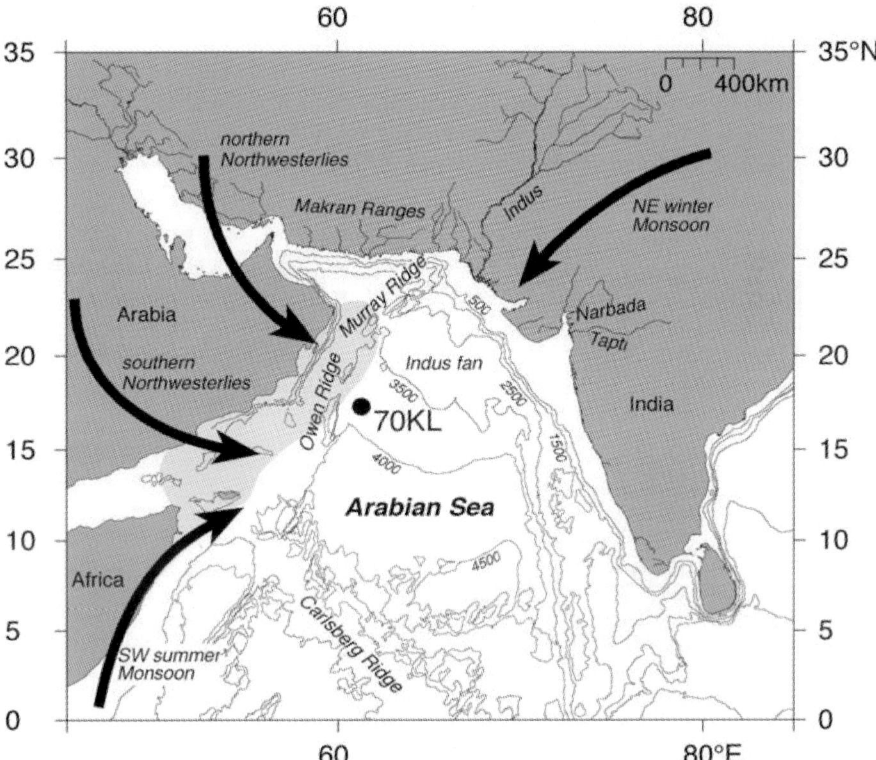

Fig. 1. Major sedimentation controlling processes in the Arabian Sea region and in particular at the coring site of sediment core 70KL (17°30'N; 61°30'E; 3810 m water depth). Aeolian input is derived from the indicated pathways of the different wind systems, the southwest summer monsoon, the northeast winter monsoon and the Arabian northwesterly winds. Significant river discharge is restricted to the Indian coast. The shaded area indicates the region of enhanced biogenic surface water productivity due to summer monsoon induced upwelling

m) after crossing the coastline of Oman. Over the Arabian Sea they overlay the southwest summer monsoon in the mid-troposphere and release their dust load (Fig. 1).

Material and Methods

The coring site of core 70KL is located in the deep Arabian Basin at the base of the Indus fan and east of the Owen Ridge at a water depth of 3810 m (Fig. 1).

The age model of core 70KL (Fig. 2) is based on eleven radiocarbon ages up to 42 ka and on the comparison to the oxygen isotope record to the global SPECMAP-stack dating back before that time (for details see Leuschner and Sirocko, 2000). Precise time markers are given by the occurrence of the Toba Ash (ca. 71 ka; Ninkovich *et al.*, 1978; Zielinski *et al.*, 1996) and the extinction of *Globigerinoides ruber pink* (about 120 ka; Thompson and Bé, 1979). Further improvement was achieved to the age model by an orbitally tuning process in the section between the fixed markers at 70 and 120 ka, where the SPECMAP tuning is ambiguous (Fig. 2; Leuschner and Sirocko, subm.).

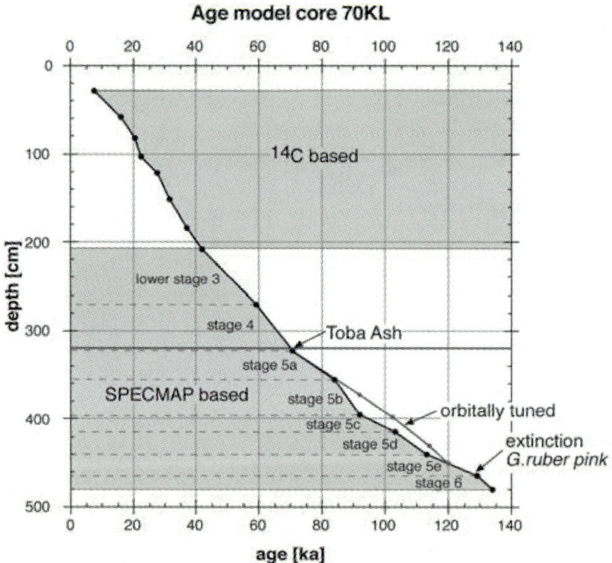

Fig. 2. The age model of core 70KL. Based on radiocarbon ages and the comparison of the oxygen isotope record with the global SPECMAP-stack (see also Leuschner and Sirocko, 2000). Radiocarbon ages include a 400 year seawater correction and calibration by means of a factor f=1.168 up to 25 ka (after Bard pers. communication at EFS-Conference, Aquifredda, 1997) and beyond after Voelker *et al.*, (1998). Additional Toba Ash (ca. 71 ka) and *Globigerinoides ruber pink* (about 120 ka) time markers. The orbital tuning between the fix markers at 70 and 120 ka is indicated by the grey line (Leuschner and Sirocko, subm.)

For the geochemical analysis core 70KL was sampled in intervals of 1 cm. Sample aliquots of 250 mg were digested in teflon vessels using a HNO_3-$HClO_4$-HF mixture. The concentrations of major elements (Na, Mg, Al, P, S, K, Ca, Ti, Mn and Fe) as well as trace elements (Li, Sc, Zn, Sr, Y, Ba and La) were measured with an ICP-AES at the GFZ-Potsdam. Using the same solutions the concentrations of Cr, Co, Ni, Cu, Zn, Ga, Rb, Zr, Mo, Cs, Ba, Hf, Pb, Th and U and the rare earth elements (REE) were measured with an upgraded VG PlasmaQuad PQ1 mass spectrometer at the ICPMS laboratory of the Institute of Geosciences, Universität Kiel. Measurements were carried out according to the procedure described in Garbe-Schönberg (1993). External precision was better than 5 %rel for most elements, and accuracy was checked with international rock standards.

The measurement of the inorganic carbon was carried out with a Ströhlein Coulomat 702-S0/CS/E at the GFZ Potsdam and the organic carbon content was measured with a Eltra METALYT-CS-1000-S at the Alfred-Wegener-Institut (AWI) in Potsdam after removing the inorganic carbonate content of the samples using heated $HCl_{conc.}$.

Several multivariate r-mode principle component analysis runs were obtained from the geochemical data using a principle component (PC) option with a correlation matrix in combination with a subsequent (varimatrix) rotation of the components. The boundary preferences for each run were set to a maximum of 10 principle components, and a tolerance of 0.001 or a maximum of 25 iterations.

Time series analyses were performed with the public domain software "AnalySeries 1.1" (Paillard *et al.*, 1996). In standard we used the Blackman-Tuckey method with a "compromise" resolution/confidence of a band width (number of lags) of 30% of the series. For statistical confidence control the time series analyses were also performed using other resolution/confidence ratios.

Geochemical principle component analysis of core 70KL and its environmental interpretation

Principle component analysis was performed on the geochemical data of core 70KL in order to obtain end-member compositions reflecting the principle processes affecting the sediment composition. At first we applied a bulk sediment analysis on 49 variables, including all 43 measured elements, 3 replicate element records, inorganic- and organic carbon, and magnetic susceptibility. The same parameters were used for a second analysis on a recalculated carbonate free basis (cfb) of the sediment.

Bulk sediment analysis

In the bulk sediment study five major components were obtained, which account for 86.7% of the total variance (Fig. 3). The most prominent factor is a lithogenic/carbonate principle component (PC 1) which accounts for 54.8% of the total variance. This component is characterised by high positive loadings on lithogenic elements such as Al, K, Ti, Fe, light REE (Ce, Pr, Nd, Sm), etc. and high negative loadings on CO_3, Ca, and Sr, reflecting the dilution effect of lithogenic input to the carbonate content and vice versa.

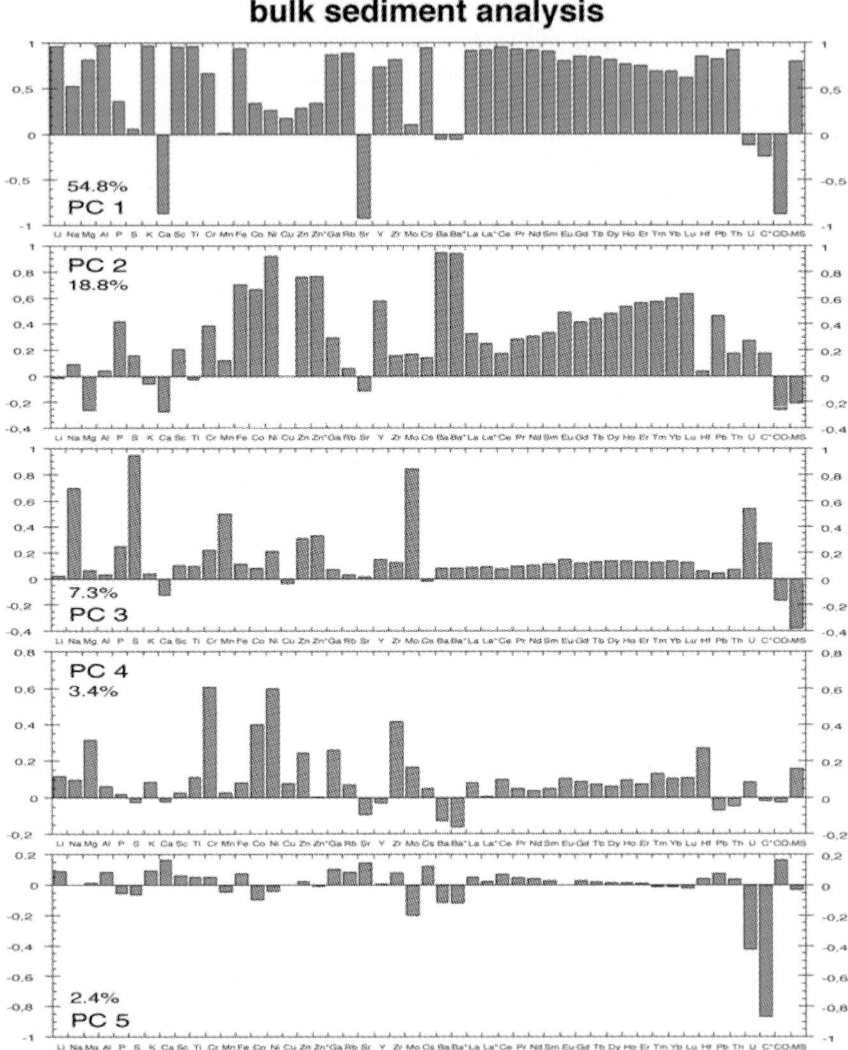

Fig. 3. Principal component factor loadings for the bulk sediment analysis. C* = organic carbon; MS = magnetic susceptibility. The Elements Zn*, Li* and Ba* were measured at the GFZ Potsdam, whereas the measurements of Zn, Li and Ba were carried out at the University Kiel

The second component (18.8% of variance) comprises high loadings on the upwelling related elements Ba, Cu, Zn, Co, Ni and moderate loadings on the heavy REE (Lu, Yb, Tm, Er, Ho). These elements reveal a close correlation to dissolved nutrients in sea water (German and Elderfield, 1990; Sunda, 1994; Löscher, 1999) and are enriched in the sediment and deep water under high productive areas while being depleted in the surface water. In such regions they are mediated to biological components by scavenging, uptake of phytoplankton and assimilation by zooplankton (Elderfield, 1990; Bruland *et al.*, 1991; Hutchins and Bruland, 1995; Jannasch *et al.*, 1996).

The third factor (PC 3; 7.3% of variance) provides high loadings on S, Mo, Mn, U and Na, as well as significant negative loading on the magnetic susceptibility. These elements are sensitive to deep ocean ventilation and/or diagenetically mobile (Shaw *et al.*, 1990). S and Mo concentrations in the seawater vary in relation to salinity (Sunda, 1994) which is confirmed by a high loading of Na. Mn is strongly affected by reductive dissolution (Sunda, 1994) and magnetic particles dissolve under reducing conditions. Therefore this factor mirrors the deep water ventilation and redox-conditions in the sediment surface.

Moderate factor loadings on Cr, Ni, Co and Zr together with minor loadings on Mg, Hf and Ga characterise the principle component 4 (PC 4; 3.4% of variance). Most of these elements were documented to be related to the fine grained lithogenic fraction at the Oman Margin (Shimmield *et al.*, 1990; Pedersen *et al.*, 1992) and the Indian shelf (Paropkari, 1990). The spatial distribution patterns of these elements (Sirocko, 1995) match the distribution of chlorite in the northern Arabian Sea surface sediments with high abundance's at the coast of Oman and Pakistan (Kolla *et al.*, 1981; Sirocko and Lange, 1991). In this part of the Arabian Sea these elements, just like the chlorite, can be attributed to originate from ophiolithic sources in Iran, Pakistan and Oman and are transported by the northern branch of the northwesterly winds from the Persian Gulf region.

The organic carbon factor (PC 5; 2.4% of variance) is dominated by a high (negative) load on organic carbon and moderate (negative) load on Mo and reflect productivity and preservation of organic matter in the sediment (Ittekkot and Arain, 1986; Pedersen *et al.*, 1992; Calvert *et al.*, 1995).

Carbonate free analysis

In the carbonate free case study eight principle components were obtained which are significant for at least a minimum of one element of the data base (Fig. 4). The eight principle components account for 88.4 percent of the total variance, but one has to be cautious to take the relative importance of the individual factors as representative for the relative importance of the associated process to the sediment composition. The values are strongly influenced by the number of covarying elements put into the analysis. In particular the high proportion of the total variance explained by the carbonate free principle component two (cfb PC2) is strongly bi-

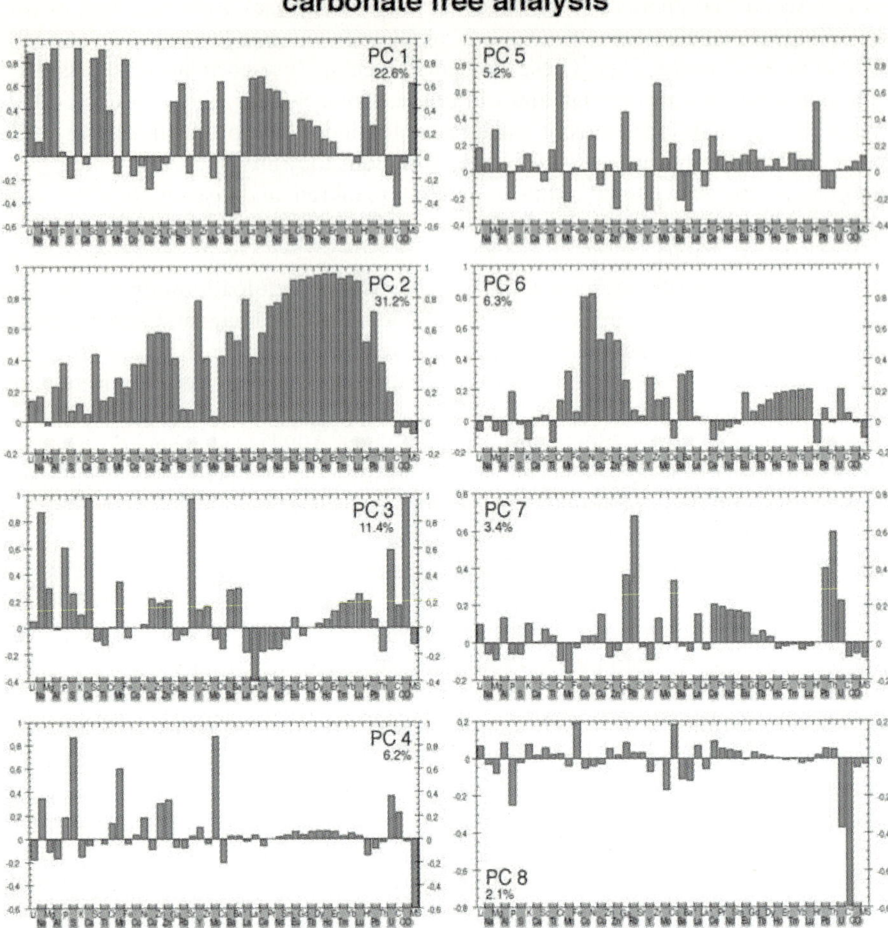

Fig. 4. Factor loadings of the principal component analysis for the same elements as in the bulk sediment analysis on a carbonate-free-base (cfb)

ased by the large number of rare earth elements in the data matrix. Four of the eight principle components revealed by this carbonate free study are comparable to components already obtained by the bulk sediment study (PC 2, PC 3, PC 4 and PC 5). The other four components reveal a more detailed picture of the principal processes contributing to the lithogenic/carbonate component (PC 1) by dividing this factor into four additional components.

The lithic component (cfb PC 1; 22.6% of the total variance)

The first principle component comprises the lithogenic elements with especially high loadings on K, Al, Ti, Li, Sc, Fe and Mg. All of these are major components in silicates. The contribution of silicates, preferentially clay minerals, to the western Arabian sea is commonly of aeolian nature. Variations of this factor should therefore reflect the aeolian input by northwesterly winds from Central Arabia, which is the most important dust source in this region (Sirocko and Sarnthein, 1989; Sirocko, 1995).

The rare earth element (REE) component (cfb PC 2; 31.2%)

High loadings on rare earth elements, particularly the heavy rare earth elements, dominate the second factor of the carbonate free study. The relative importance of this factor is due to the relatively large number of these elements with similar behaviour during scavenging and deposition.

To investigate the principal processes being responsible for the REE supply into the Arabian Sea sediments another principle component analysis was performed including the measured REE concentrations and the concentration of the neighbouring element Hf (in the periodic table). The study reveal two major factors for the downcore variance of the REE concentrations that account for 90.0% of the variance in the data (Fig. 5a).

The first component (REE PC 1; 48.9%) is highly loaded on the heavy rare earth elements (HREE) Lu, Yb, Tm, Er and Ho. The light rare earth elements (LREE) Ce, Pr, Nd and Sm together with La show high loadings on the second component (REE PC 2; 41.1%), whereas the intermediate REE exhibit moderate contributions on both of the two factors. A strong correlation ($r=0.81$) between the LREE factor scores and the Al content variations in core 70KL indicates that a part of the REE variation is due to lithogenic discharge (Fig. 5b). As the Al content mainly reflects the aeolian input via northwesterly winds from central Arabia the LREE factor may serve as a proxy for this particular transport process, too.

In the Arabian Sea there is generally a depletion of dissolved REE concentrations in the surface waters with a relative enrichment in the oxygen minimum zone and an increase with further depth (German and Elderfield, 1990). This profile in the water column parallels the nutrient content. Such nutrient like behaviour of the REE is attributed to a biologically mediated uptake or scavenging of the REE and a release deeper within the water column. In core 70KL this behaviour of the REE, in this case particularly the HREE, is confirmed by the covariance of the REE PC 1 with the Ba/AL-record as well as the upwelling factors PC 2 and cfb PC 6. The latter factors contain high loadings on trace metals which are known to be scavenged by organic particles (Fig. 5b).

The carbonate component (cfb PC 3; 11.4%)

Factor loadings of the carbonate component are high on CO_3, Ca and Sr, but also on Na and moderate on U and P. These elements are commonly related to biogenic matter and the concentration in the sediment is dominantly coupled to primary productivity and dilution by lithogenic influx.

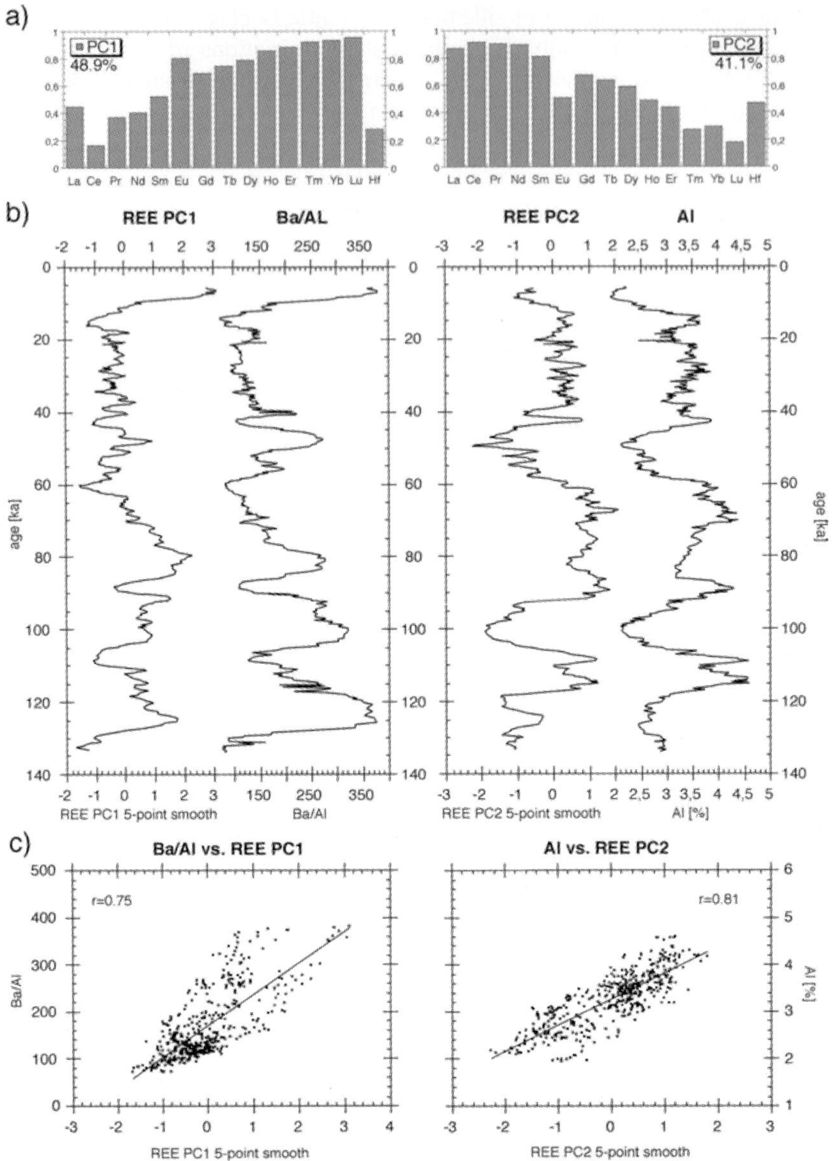

Fig. 5. a) Factor loadings of the two rare earth element (REE) principal components. Factor 1 shows high loadings (>0.8) on the heavy REE Lu, Yb, Tm, Er and Ho. The light REE Ce, Pr, Nd, Sm and La exhibit high loadings on factor 2, whereas the intermediate elements Eu, Gd, Tb and Dy show moderate loadings (0.5 -0.8) on both of the components.
b) Downcore variations and comparison of the two rare earth element components compared with the Ba/Al upwelling record and the Al (terrigenous) record in core 70KL.
c) The x/y comparison shows good correlation between the REE PC1 and the Ba/Al productivity record (r=0.75) and between the REE PC2 and the clastic Al-record (r=0.81).

The redox component (cfb PC 4; 6.2%)

This component summarizes the elements S, Mn and Mo and is comparable to the bulk sediment component PC 3.

The ophiolothic component (cfb PC 5; 5.2%)

The ophiolithic component shows high loadings on the elements Cr, Zr and Hf. This component is similar to the bulk sediment component PC 4.

The upwelling component (cfb PC 6; 6.3%)

With high loadings on Ni, Co, Zn, Cu and Ba this factor is indicative for the upwelling intensity and has a counterpart in the bulk sediment study PC 2.

The rubidium component (cfb PC 7: 3.4%)

This component is dominated by high loadings on Rb and Th as well as weak to moderate loadings on Pb, Ga and Cs. Most of these elements show high concentrations along the Pakistani and Indian coast (Sirocko, 1995), thus implying an Indian source. Concentrations of Rb and Cs are high in the northern Arabian Sea and at the mouth of the rivers Indus, Narbada and Tapti, whereas the Pb concentrations are highest at the southern Indian coast. It is unknown whether variations in these elements are the recording of changing river discharge by the Indus or if they are due to changes in the aeolian dust transport by the northeast winter monsoon.

The organic carbon component (cfb PC 8; 2.1%)

This organic carbon component with a high (negative) loading on the organic carbon content is almost equal to the PC 5 in the bulk sediment study.

Relative variations of dust transport and monsoon associated upwelling during the last glacial cycle

Clastic input to the site of core 70KL mainly derived from the northwesterly winds from the central Arabian deserts and the Persian Gulf region. Dust coming from the latter source (cfb PC 5, Fig. 6a) oscillates continuously during the last glacial cycle, but with no systematic variation regarding climatic changes as indicated by the oxygen isotope record. The amount of clastic material from central Arabia (cfb PC 1, Fig. 6a) shows a temporal variation, which is close to the global alternation of cold and warm stages in the oxygen isotope record. A strong decrease in the dust supply from central Arabian sources is commonly observed during warm stages, e.g. 5e, 5c, the early stage 3 and the Holocene.

Sediment input from India (cfb PC 7, Fig. 6a) shows peak values during cold stages. Enhanced deposition of such material should be due to increasing winter monsoon winds. Another explanation could be a stronger riverine influx into the deep-sea as a result of lower sea level, whereas the river discharge is distributed onto the shelf regions during high stands of the sea-level.

Factor scores of the REE-component, the upwelling-component and the carbonate component are all related to the southwest summer monsoon. The upwelling/scavenging-factor (cfb PC 6, Fig. 6a) probably reflects the most accurate recording of summer monsoon strength because the other two records are in addition to this significantly biased through lithogenic contributions. As demonstrated the REE-component is also positively correlated to lithogenic flux rates. In turn the carbonate record will be diluted by higher lithogenic contributions. The effect of this divergent behaviour is most significant between 48 and 56 ka where a broad maximum in the carbonate content is due to a decreased lithogenic input. However, the four individual productivity (upwelling) peaks superimposed onto this broad maximum can be observed in all records. Generally the variation of all the three records reveal higher values during warmer stages (grey marked). The contrast between warm and cold stages is most visible and strongest in the carbonate factor, because contribution of lithic components is enhanced during cold stages, whereas biogenic productivity is strongest during warm phases.

Time series analysis

In order to evaluate the driving mechanisms behind the millennial scale variations in the Indian Monsoon system a Blackman-Tuckey time series analysis was carried out on the four principle components related to the main wind directions of the region (Fig. 6b). The upwelling component cfb PC 6 exhibits five frequencies considered significant during the entire last glacial cycle, i. e. 5600, 4400, 2000, 1650 and 1200 years. All of these were close to previously documented cycles in the Indian Summer Monsoon or associated upwelling. Corresponding periods to the 5600 year and 4400 year period are given by 5800 and 4400 year oscillations in the *Globigerinoides ruber* abundance in the Arabian Sea during the last glacial (Pestiaux *et al.*, 1988). Naidu and Malmgren (1995) observed a 2200 year oscillation and Sirocko et al. (1996) observed variations of 1785 and 1150 years in the abundance of *Globigerina bulloides* during the last 20.000 years.

The northeast monsoon shows periodic variations on 3500, 2000, 1200 and 950 years. The latter one is the dominating frequency in the Rb/Al-ratio in core 74KL during the last 20.000 years, which is attributed to record variations in the northeast monsoon (Sirocko *et al.*, 1996).

Cyclic oscillations in the Arabian northwesterly winds (northern branch; cfb PC 5) are dominated by periods of 3200, 2000, 1450 and about 1050 years. Most striking is the presence of the 1450 year cycle throughout the entire last glacial

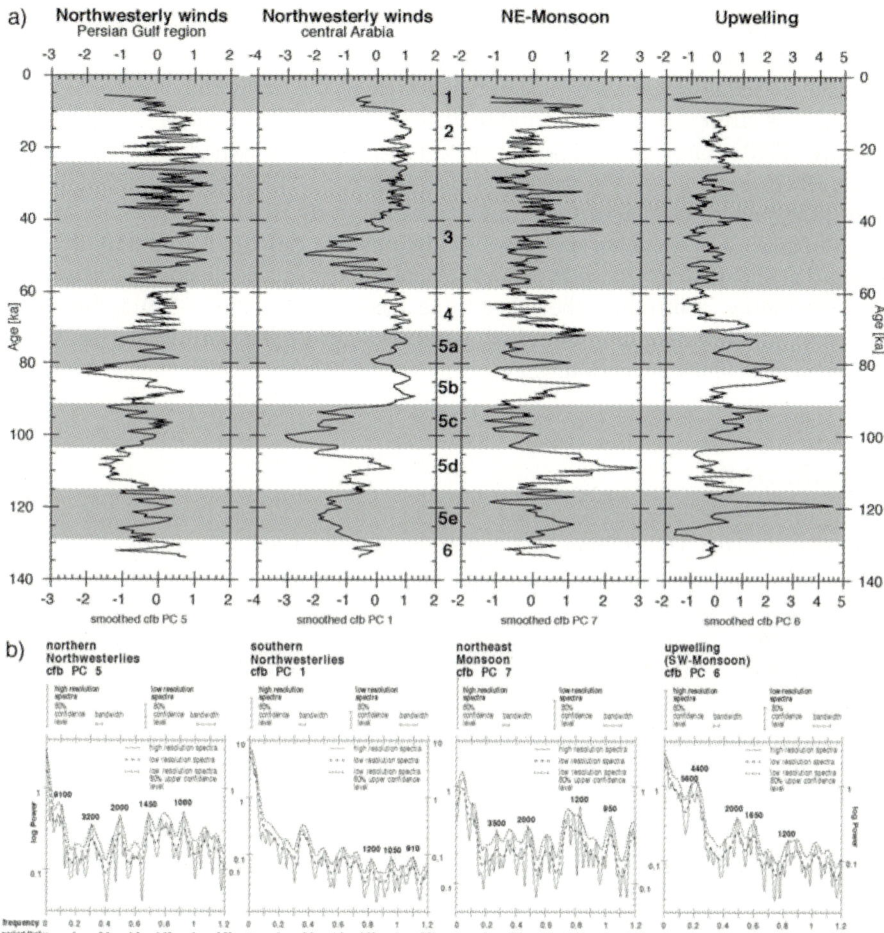

Fig. 6. a) Time series of four principal components from the geochemical analysis. These are indicative for the lithogenic input derived from the northeast monsoon (PC 7), the northwesterly winds (southern branch PC 1; northern branch PC 5) and the "organic matter component" (PC 6). The latter is probably the most independent indicator for the upwelling intensity and therefore for the strength of the southwest Monsoon. All records are smoothed by the use of a 5 point running mean. Numbers in the middle indicate the oxygen isotope stages with warmer stages marked in grey.

b) Blackman-Tuckey spectral analysis on the prewhitened (f=0.5) records of upwelling (cfb PC 6), northeast Monsoon (cfb PC 7), southern Northwesterly winds (cfb PC 1) and northern Northwesterly winds (cfb PC 5). The records were resampled for analysis on an Δt=270 years. Analyses were performed on a high resolution (193 lags or 30% of series; bandwidth=0.039) and low resolution (64 lags or 10% of series; bandwidth=0.117) mode. 80% confidence level and bandwidth bars are shown above. Frequencies were considered significant when the high resolution spectra (solid line) passes the upper 80% confidence level of the low resolution spectra (dashed line).

cycle. This period is also present in the dolomite, Mn and light REE factor record of core 74KL during the last 20.000 years (Sirocko *et al.*, 1996). A similar oscillation is even the dominating period in the climate of the northern hemisphere as indicated by a strong 1470 year period in the oxygen isotope record of the Greenland ice core GISP2 (Grootes and Stuiver, 1997) and in north Atlantic sediments (Bond *et al.*, 1997).

In the southern branch of the Arabian northwesterly winds periodicities are not as significant as in the other records. Oscillations of 1200, 1050, and 910 years indicate that the aeolian transport from the central Arabian peninsular is influenced by periodicities controlling the other three records. As there are humidity, controlled by the summer monsoon, and wind strength of the northwesterlies themselves on the 1050 year period.

Conclusion

The geochemistry of Arabian Sea sediments can be used to investigate relative variations of the sediment transport capacity with the Indian Monsoon system. After statistical analysis of the sediment's chemical composition several principal components of the sediment's chemical composition can be correlated to the principal environmental processes of the region. Changing lithogenic influx derived from the major wind systems of the region (the Arabian northwesterly winds, the northeast winter monsoon and the southwest summer monsoon) associated upwelling and biogenic productivity as well as the redox conditions at the sediment-water interface are the most important processes controlling the sedimentation in the central Arabian Sea. This study has shown that at least the aeolian input from the different source areas derived from the seasonally changing winds is controlled by periodicities. These were significant and persistent during the entire last glacial cycle. In particular the observation of the periodicity at 1450 years in the amount of dust from the Persian Gulf area shows that climatic connections between the high northern latitudes and the subtropical monsoon climate were present throughout the Late Pleistocene.

References

Behl, R.J., and Kennett, J.P. (1996) Brief interstadial events in the Santa Barbara basin, NE Pacific, during the past 60 kyr. *Nature*, **379,** 243-246.

Bond, G.C., Showers, W., Cheseby, M., Lotti, R., Almasi, P., deMenocal, P., Priore, P., Cullen, H., Hajdas, I., and Bonani, G. (1997) A pervasive millennial-scale cycle in North Atlantic Holocene and glacial climates. *Science*, **278,** 1257-1266.

Bruland, K.W., Donat, J.R., and Hutchins, D.A. (1991) Interactive influences of bioactive trace metals on biological production in oceanic waters. *Limnology and Oceanography*, **36,** 1555-1577.

Calvert, S.E., Pedersen, T.F., Naidu, P.D., and von Stackelberg, U. (1995) On the organic carbon maximum on the continental slope of the eastern Arabian Sea. *Journal of Marine Research*, **53,** 269-296.

Elderfield, H. (1990) Tracers of ocean paleoproductivity and paleochemistry: an introduction. *Paleoceanography*, **5,** 711-717.

Findlater, J. (1969) A major low-level air current near the Indian Ocean during the northern summer. *Q.J.R. Meterol. Soc.*, **95,** 362-380.

Garbe-Schönberg, D. (1993) Simultaneous determination of thirty-seven trace elements in twenty-eight international rock standards by ICP-MS. *Geostandard Newsletter*, **17,** 81-97.

German, C.R., and Elderfield, H. (1990) Rare earth elements in the NW Indian Ocean. *Geochimica et Cosmochimica Acta*, **54,** 1929-1940.

Grootes, P.M., and Stuiver, M. (1997) Oxygen 18/16 variability in Greenland snow and ice with 10^{-3} to 10^{5}-year time resolution. *Journal of Geophysical Research*, **102,** 26,455-26,470.

Hutchins, D.A., and Bruland, K.W. (1995) Fe, Zn, Mn, and N transfer between size classes in a coastal phytoplankton community: trace metal and major nutrient compared. *Journal of Marine Research*, **53,** 297-313.

Ittekkot, V., and Arain, R. (1986) Nature of particulate organic matter in the river Indus, Pakistan. *Geochimica et Cosmochimica Acta*, **50,** 1643-1653.

Jannasch, H.W., Honeyman, B.D., and Murray, J.W. (1996) Marine scavenging: the relative importance of mass transfer and reaction rates. *Limnology and Oceanography*, **41,** 82-88.

Kolla, V., Kostecki, J.A., Robinson, F., and Biscaye, P.E. (1981) Distributions and Origins of Clay Minerals and Quartz in Surface Sediments of the Arabian Sea. *Journal of Sedimentary Petrology*, **51,** 563-569.

Leuschner, D.C., and Sirocko, F. (2000) The low latitude monsoon climate during Dansgaard-Oeschger cycles and Heinrich Events. *Quaternary Science Reviews*, **19,** 243-254.

Leuschner; D.C. and Sirocko, F. (subm.) Orbital insolation forcing of the Indian Monsoon - a Motor for global climate change?. Palaeogeography, Palaeoclimatology, Palaeoecology.

Löscher, B.M. (1999) Relationships among Ni, Cu, Zn, and major nutrients in the Southern Ocean. *Marine Chemistry*, **67,** 67-102.

McGregor, G.R., and Nieuwolt, S. (1998) "Tropical climatology." Wiley & Sons, New York.

Naidu, P.D., and Malmgren, B.A. (1995) A 2,200 years periodicity in the Asian monsoon system. *Geophysical Research Letters*, **22,** 2361-2364.

Ninkovich, D., Shackleton, N.J., Abdel-Monem, A.A., Obradovich, J.D., and Izett, G. (1978) K-Ar age of the late Pleistocene eruption of Toba, north Sumatra. *Nature*, **276,** 574-577.

Paillard, D., Labeyrie, L., and Yiou, P. (1996) Macintosh program performs time-series analysis. *EOS*, **77,** 379 (http://www.ngdc.noaa.gov/paleo/softlib.html).

Paropkari, A.L. (1990) Geochemistry of sediments from the Mangalore-Cochin shelf and upper slope off southwest India: geological and environmental factors controlling dispersal of elements. *Chemical Geology*, **81,** 99-119.

Pedersen, T.F., Shimmield, G.B., and Price, N.B. (1992) Lack of enhanced preservation of organic matter in sediments under the oxygen minimum on the Oman Margin. *Geochimica et Cosmochimica Acta*, **56,** 545-551.

Pestiaux, P., van der Mersch, I., Berger, A., and Duplessy, J.C. (1988) Paleoclimatic Variability at frequencies ranging from 1 cycle per 10,000 years to 1 cycle per 1000 years: evidence for nonlinear behaviour of the climate system. *Climate Change*, **12,** 9-37.

Prins, M.A. (1999) "Pelagic, Hemipelagic and Turbidite Deposition in the Arabian Sea during the Late Quaternary." Unpublished PhD thesis, Universiteit Utrecht.

Reichart, G.J., Lourens, L.J., and Zachariasse, W.J. (1998) Temporal variability in the northern Arabian Sea Oxygen Minimum Zone (OMZ) during the last 225,000 years. *Paleoceanography*, **13,** 607-621.

Schulz, H., von Rad, U., and Erlenkeuser, H. (1998) Correlation between Arabian Sea and Greenland climate oscillations of the past 110,000 years. *Nature*, **393,** 54-57.

Shankar, R., Subbarao, K.V., and Kolla, V. (1987) Geochemistry of surface sediments from the Arabian Sea. *Marine Geology*, **76,** 253-279.

Shaw, T.J., Gieskes, J.M., and Jahnke, R.A. (1990) Early diagenesis in differing depositional environments: The response of transition metals in pore water. *Geochimica et Cosmochimica Acta*, **54,** 1233-1246.

Shimmield, G.B., Price, N.B., and Pedersen, T.F. (1990) The influence of hydrography, bathymetry and productivity on sediment type and composition of the Oman Margin and in the Northwest Arabian Sea. *In:* (A. H. F. Searle, and A. C. Ries, Eds.) *"The Geology and Tectonics of the Oman Region."*, pp. 759-769. Geological Society Special Publication.

Sirocko, F. (1995) Abrupt change in monsoonal climate: evidence fromthe geochemical composition of Arabian Sea sediments, Habilitation Thesis, 216 pp., Christian Albrechts Universität, Kiel.

Sirocko, F., Garbe-Schönberg, and D. Devey, C. (2000) Processes controlling trace element geochemistry of Arabian Sea sediments during the last 25,000 years, *Global and Planetary Change*, **26**, 217-303.

Sirocko, F., Garbe-Schönberg, C.-D., McIntyre, A., and Molfino, B. (1996) Teleconnections Between the Subtropical Monsoons and High-Latitude Climates During the Last Deglaciation. *Science*, **272,** 526-529.

Sirocko, F., and Lange, H. (1991) Clay mineral accumulation rates in the Arabian Sea during the Late Quaternary. *Marine Geology*, **97,** 105-119.

Sirocko, F., and Sarnthein, M. (1989) Wind-borne deposits in the Northwestern Indian Ocean: record of Holocene sediments versus modern satellite data. *In:* (M. Leinen, and M. Sarnthein, Eds.) *"Paleoclimatology and Paleometeorology: Modern and Past Patterns of Global Atmospheric Transport."*, pp. 401-433. Kluwer Academic Publishers, Dordrecht, Boston, London.

Sirocko, F., Sarnthein, M., Erlenkeuser, H., Lange, H., Arnold, M., and Duplessy, J.-C. (1993) Century-scale events in monsoonal climate over the past 24,000 years. *Nature*, **364,** 322-324.

Sunda, W.G. (1994) Trace Metal/Phytoplankton interactions in the Sea. *In:* (G. Bidoglio, and W. Stumm, Eds.) *"Chemestry of Aquatic Systems: Local and Global Perspectives."*, pp. 213-247. Kluwer Academic Publishers, Dordrecht.

Thompson, P.R., and Bé, A.W.H. (1979) Disappearance of pink-pigmented *Globigerinoides ruber* at 120,000 yr BP in the Indian and Pacific Oceans. *Nature*, **280,** 554-558.

Voelker, A.H.L., Sarnthein, M., Grootes, P.M., Erlenkeuser, H., Laj, C., Mazaud, A., Nadeau, M.-J., and Schleicher, M. (1998) Correlation of marine ^{14}C ages from the nordic seas with the GISP2 isotope record: implications for radiocarbon calibration beyond 25 ka BP. *Radiocarbon*, **40,** 514-534.

Zielinski, G.A., Mayewski, P.A., Meeker, L.D., Whitlow, S., Twickler, M.S., and Taylor, K. (1996) Potential atmospheric impact of the Toba mega-eruption ≈71,000 years ago. *Geophysical Research Letters*, **23,** 837-840.

Little Ice Age climatic fluctuations in the Namib Desert, Namibia, and adjacent areas: Evidence of exceptionally large floods from slack water deposits and desert soil sequences

Klaus Heine

University of Regensburg
Institute of Geography
Regensburg, Germany

Abstract

Knowledge of long-term rainfall variablity is essential for water management in Namibia. Data relevant to assess this variability are scarce because of the lack of long instrumental climate records and the limited potential of standard high-resolution proxy records. In northern and eastern Africa the reconstruction of Holocene tropical lake-level changes has established alternating phases of desiccation and of high stands with lake-levels more than 100 m above the present level. This record of paleohydrological changes is impressive as compared to available data collected from modern instrumented observations. Such sudden and dramatic changes of the hydrologic regime within time scales that are relevant to human societies are not known from southwestern arid Africa (Namib Desert). Fluvial silts, accumulated in some Namib valleys, are interpreted as records of reduced precipitation in the catchments. Our investigations show that these fluvial silts are slack water deposits (SWDs) and reflect hydrologic - and climatic - conditions during the late Holocene that caused extreme flash floods in the valleys. Here we describe SWDs of some Namibian Desert valleys and present ^{14}C dates of their ages. The youngest accumulation phase occurred during the Little Ice Age (LIA)(ca. AD 1300 to 1850). The biggest flash floods of the LIA, in most catchments, experienced water levels in the valleys that exceeded the most extreme floods of the last 100 - 150 years. In the northwestern Namib Desert, flash floods of the LIA were more frequent and more extreme than in the central Namib Desert. This may be caused by small shifts of the tropical-temperate-troughs in southern Africa and the south west Indian Ocean.

1 Introduction

In Europe, the general turn towards colder climates from AD 1200 - 1400 onwards, accompanied by shifts of the zones of most cyclonic activity as the polar cap and the circumpolar vortex expanded, and which in the seventienth century seems to have produced a world-wide cold stage, is widely known as the climatic worsening of the Late Middle Ages (LAMB 1977:449). This period of glacial advance of the last few centuries is also known as the *Little Ice Age* (LIA) because not only in Europe but in most parts of the world the extent of snow and ice on land and sea seems to have attained a maximum as great as, or in most cases greater than, at any time since the last major ice age (GROVE 1988; LAMB 1977:461f.). Since there were regional variations to this climatic deterioration, it is difficult to define a universally applicable date for the onset and the end of this period (BRADLEY 2000). Within the period AD 1550 - 1850 there was a considerable temperature variation both in time and space. The complexity (or structure) in the climate of the LIA is a reflection of the wealth of information from paleoclimate archives. The LIA was undoubtedly one of the coldest intervals in the entire Holocene (BRADLEY 2000). A consideration of possible causes of the LIA presents many different processes (GROVE 1988). The climatic phase of the LIA is characterised by - on average - reduced sun spot activity (HUPFER et al. 1998). There is evidence that the so-called Maunder minimum (the end of the Maunder minimum is dated AD 1645 to AD 1715), as a result of alterations of the solar magnetic field, may have influenced the global climate. Furthermore, volcanic eruptions may have also affected the climate during LIA times (HUPFER et al. 1998). At present, it is difficult to unequivocally ascribe the LIA climate changes to external forcing (solar, orbital, volcanic, CHAMBERS et al. 1999), to internal ocean-atmosphere interactions, or to a combination of all these, perhaps varying in importance over time (BRADLEY 2000).

Little is known about the consequences of the LIA climatic fluctuations in southwestern arid Africa. To date only RUST (1997), VOGEL & RUST (1987, 1990) and HEINE et al. (2000) mention LIA fluvial sediments and provide paleoclimatic interpretations.

In Namibia it is difficult to date sediments and soils of the LIA period. In this arid environment hardly any pollen sequences exist (SCOTT et al. 1991), nor are there trees suitable for tree ring chronologies, or laminated lake sediments, Holocene cave sinter and the like. Furthermore, only a few dating methods are available. Radiocarbon dating of organic material is limited by calibration problems of

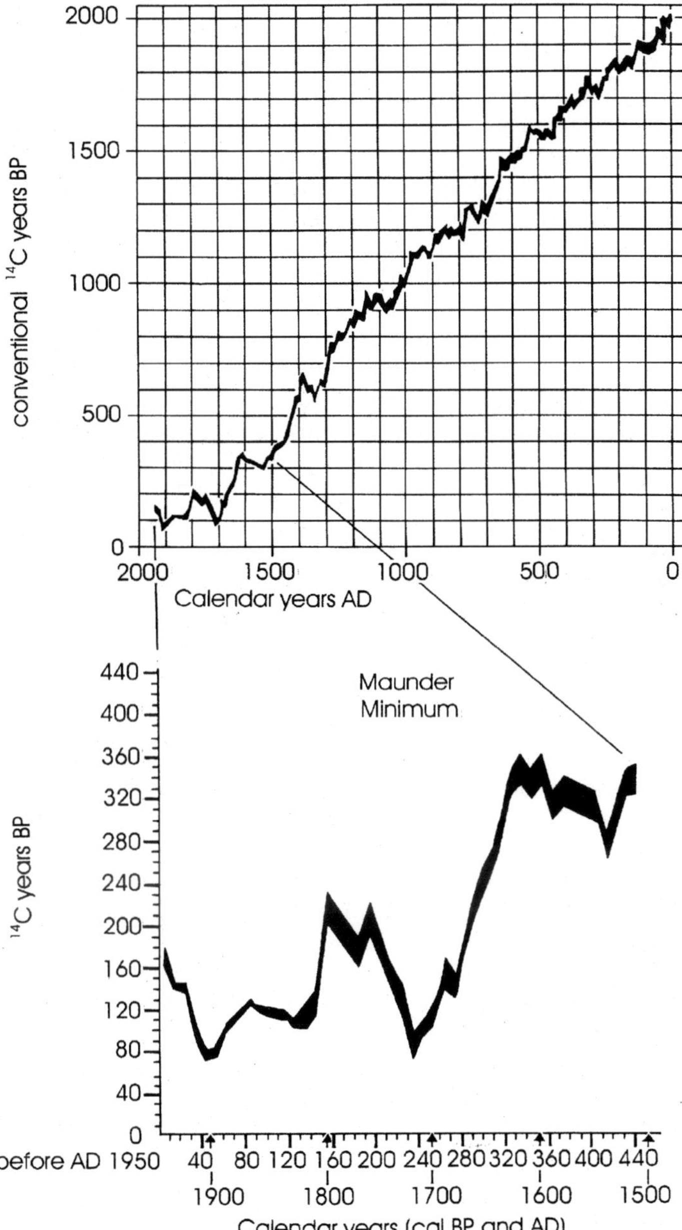

Fig. 1. The relationship between ^{14}C years and calendar years calibrated from tree rings. The width of the curve, marked in black, is twice the standard deviation given by the laboratory. Any one radiocarbon year is equivalent to more than one calendar year. Thus a radiocarbon age of 220+/-50 years is equivalent to all the calendar dates with the intervals AD 120-210, 280-320 and 410-420 (from STUIVER 1978 in GROVE 1988)

^{14}C values with calendar ages (Fig. 1). Within the last 500 years it is not possible to obtain an unambiguous calendar age from a single ^{14}C date. ^{14}C age determinations cannot be used for identifying synchroneity of LIA events. They can only provide a rough age estimation during LIA times. Thermoluminescence dating (TL) and optically stimulated luminescence dating (OSL) have not been applied to LIA sediments in Namibia, apart from a few examples of neoglacial age (BUCH et al. 1992; EITEL et al. 1999). Unlike the situation in Europe, where many historical data give evidence about the beginning of the LIA climatic fluctuation ca. 1300 AD and about the weather conditions during the seasons of the year as well as about extreme climatic hazards during the LIA period (PFISTER 1999), in Namibia historical records are short or entirely absent.

Here we present new data from fluvial sediments of some Namibian Desert valleys that document extreme precipitation events (flash floods) during the LIA period. The fluvial silty accumulations are regarded as *slack water deposits* (SWDs) of floods of high magnitude, thus representing extreme precipitation events in the upper reaches of the river catchments. This interpretation is in contrast to previous work (e.g. VOGEL 1982, 1989; RUST 1997; VOGEL & RUST 1987, 1990; EITEL et al. 1998, 1999, 2000; BLÜMEL et al. 2000a, 2000b). Paleoflood hydrology can extend the temporal and spatial flood record and thereby complement and improve the accuracy of flood prediction (ZAWADA 1997).

2 Area of investigation

In southwestern Africa the Namib Desert stretches over 2000 km along the Atlantic coast from the Olifants River (32°S) in the south to the Carunjamba River (14°S) in the North. The width of the extremely arid Namib Desert varies from 40 to 120 km. In the east, the Namib Desert reaches to the Great Escarpment. The area of the Namib Desert is situated in a region of relative tectonic stability. Different authors give various dates for the age of the Namib Desert (see HEINE & HEINE, 2002). The Benguela Current causes the extreme aridity of the Namib Desert, at least since the late Miocene (6.7 Ma), with very vigorous upwelling developing in the late Pliocene (ca. 2.2 Ma) (ROGERS 2000). During Quaternary times the Namib Desert was arid to extremely arid. Whether the coastal region of the Namib Desert was influenced by strong hygric fluctuations during Quaternary times is the subject of much debate (HEINE 1998a, 1998b; PARTRIDGE et al. 1999).

According to LANCASTER (1989), the Namib Desert can be subdivided into four main areas: the southern or transitional Namib, which includes coastal Namaqualand and the Sperrgebiet; the Namib Sand Sea; the central Namib Plains; as well as the northern Namib and Skeleton Coast (Fig. 2).

The climate of the Namib is arid to hyper-arid and especially in coastal areas is relatively cool. To the north, there is the summer rainfall desert of Angola, whilst to the south, rainfall occurs mostly in winter. A major feature of the Namib Desert

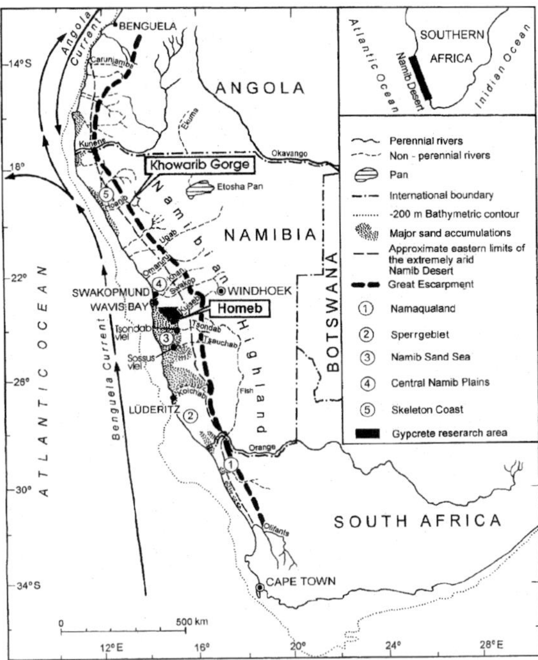

Fig. 2. The Namib Desert and adjacent areas. Map with sand accumulations and major drainage systems

is the steep climatic gradient from the cool, foggy hyper-arid coastal zone to the hotter inland areas towards the Great Escarpment, which receives summer rainfall of about 100 mm/a (LANCASTER 1989). The extension of the arid desert results from the latitudinal position of the region and the dominant effects of subtropical anticyclonic cells, especially that situated over the South Atlantic Ocean at 30° in summer. In the central Namib, moist air masses derived from the Indian Ocean can penetrate the desert only when this anticyclonic cell is weak (TYSON 1986). The moist air must cross the subcontinent to reach the Namib Desert. Thus descending divergent air masses tend to occur all year. The effects of the subsistence induced stability are reinforced by the presence of the cold Benguela Current offshore, which intensifies the temperature inversion (LANCASTER 1989).

Rainfall is highly variable and localised spotty. The nature of rainfall variability over southern Africa has been investigated on numerous timescales (synoptic, interannual, decadal and millennial, see TODD & WASHINGTON 1999). Mean annual rainfall increases from 10-15 mm at the coast to 80-100 mm near the Great Escarpment and to about 300-400 mm in the upper reaches of the Namib river catchments. From north to south, the annual total and proportion of winter rainfall increases. The circulation pattern of the central and southern Namib are strongly influenced by the South Atlantic anticyclone. Superimposed on the pattern of stable outblowing winds from the South Atlantic high pressure cell are the effects of local topographically and thermally induced circulations (LINDESAY & TYSON

1990). The overall strength of the southerly to southwesterly winds decreases from south to north and from the coast inland. It is at its weakest during the winter months (LANCASTER 1989).

Twelve major ephemeral rivers flow through the central and northern Namib Desert (Fig. 2). A number of smaller rivers originate in the arid coastlands. The geological, geographical, climatic and biological features make each catchment unique. The frequency of flooding in the rivers varies from river to river and is related to catchment size, rainfall intensity and distribution, and - recently - in some areas, upstream dams (JACOBSON et al. 1995). Floods in the Namib Desert are often sudden and devastating in their magnitude and unexpectedness (JACOBSON et al. 1995; KINGHORN 2000). There is little chance to predict their spatial and temporal occurrence.

3 Methods

Sediments deposited in the backwaters of large floods may accumulate thick sequences in tributary mouths. SWDs represent the most accurate paleoflood evidence for reconstructing the magnitude and recurrence frequency of floods that are hundreds to thousands of years old (ZAWADA 1997). Stratigraphic and sedimentologic studies of SWDs sequences combined with radiocarbon dating is used to establish a Holocene paleoflood record for some Namib Desert valleys. We mapped SWDs, which are usually fine grained and deposited from heavily sediment-laden flood waters, at sites that experience sudden reductions of flow regime. Each successive flood with a stage capable of inundating a previously accumulated slack water sediment will deposit a new layer on top of the previous one. Smaller floods will deposit sediment as insets that exhibit an onlapping relationship with the existing SWDs. The maximum elevation of the SWDs is assumed to indicate the peak stage of the flood (ZAWADA 1997). The value of SWDs as proxy indicators of paleoflow is determined by the extent to which they are preserved. It is recognised that erosion of SWDs can lead to gaps in the sedimentary record.

We analysed the grain size distribution, carbonate content and clay mineral associations to discern between different paleoflood sediments of different origins. Sedimentary structures usually show flat lamination, implying slack water sedimentation by moderate rates of deposition, as suggested by ZAWADA (1997) for South Africa, rather than sudden or rapid rates as documented for the lower Pecos River in Texas/USA by KOCHEL & BAKER (1982).

Detailed studies of many sections of the soils of the central Namib Desert in the area between the Swakop and the Kuiseb valleys (HEINE & WALTER 1996) supplement the analyses carried out on the SWDs and allow reconstructions of extreme precipitation events (sheet wash).

In the laboratory the following analyses were carried out: Grain size by standard techniques (sieve, pipette), $CaCO_3$ content by the Scheibler & Finkmer technique, % organic material by using UV-VIS spectrometer (Lambda 2), colour

(MUNSELL), clay mineral identification by X-ray diffraction (Siemens X-ray unit D5000), ^{14}C age determination of leaves, twigs, charcoal etc. (^{14}C laboratory in Hannover and AMS ^{14}C in Erlangen).

4 The Hoanib River, Kaokoveld, northern Namib Desert

The Hoanib catchment comprises 17,200 km^2 between the Atlantic Ocean and the Otjovasandu area in the east (Fig. 3). The river length is 270 km and the catchment elevation range is from 0 to 1821 m a.s.l. In this area the Great Escarpment has not been developed because tectonic processes dominated macroform development (BRUNOTTE & SPÖNEMANN 1997). Rupture fissures, fault scarps, up- and downwarpings have created a mainly rift-parallel pattern of landforms. Epigenetic reaches of the Hoanib valley form deep and spectacular gorges such as the *Khowarib Schlucht* (Khowarib Gorge). Mean annual precipitation is less than 20 mm at the coast and about 325 mm in the east.

A schematic morphological sketch of the valley structure from Kaokoveld that can be applied to the Hoanib catchment is contained in LESER (2000) (Fig. 4). SWDs occur in many sites along the valley. The deposits of the Khowarib Gorge, of the basin areas east and west of the Naweb Gorge, and of the Hoanib valley east of the Tsuxub River (Amspoorts Silts) were included into our investigations.

In the Khowarib Gorge, Holocene SWDs accumulated at various sites (Fig. 5). In the east, at the entrance of the gorge, brown (7.5YR5/4) SWDs (member 2) were deposited on greyish, bedded, weakly cemented, fluvial sands (sand: 45-50%; silt: 33-38%; clay: 16-22%; colour: 2.5Y7/4, 10YR5/4 to 10YR8/1). These brown SWDs consist of >50% fine sand. The brown SWDs filled the gullies that are incised into the greyish sediments. After deposition, the brown SWDs were eroded in many places, so that only remnants are to be found between the older greyish deposits. In places where the brown SWDs were not eroded, they are slightly cemented surficially. Concretions of cemented sediments are found mainly near the top of the sequence. A few km downvalley, the brown SWDs only occur in small remnants in the gully relief of the greyish deposits. The height of the accumulation of the brown SWDs did not reach to the top of the older greyish fluvial deposits. Furthermore, in many places slope debris accumulation covered the greyish sediments. The debris consists of coarse angular blocks that were transported through steep tributary valleys and ravines into the main valley. The slope debris was transported by running water. The sections show that the debris accumutated on top of the greyish deposits prior to the erosion of the gullies and prior to the deposition of the brown SWDs.

The weakly cemented brown silty sediments (SWD member 2) of the upper Khowarib Gorge were TL-dated by EITEL et al. (2000) to 9.01+/-1.2 ka (IR-OSL data by B. MAUZ). This age is confirmed by the weak pedogenetic processes as well as the gully development.

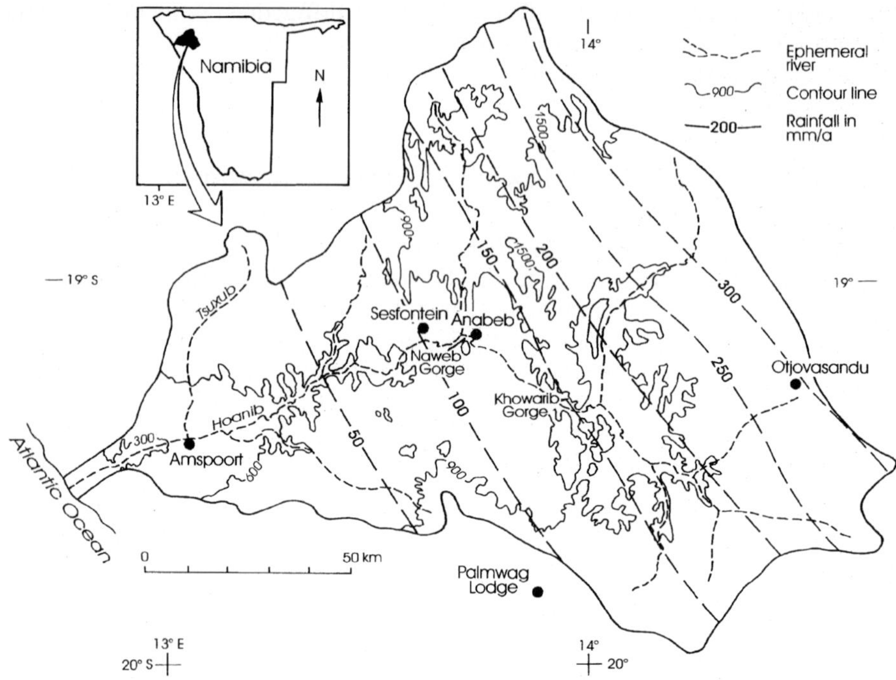

Fig. 3. The Hoanib catchment (after JACOBSON et al. 1995)

At 19°19'31''S/13°59'11''E (Fig. 5) some sections document the accumulation and erosional phases of the Hoanib Valley in the most complete way. Red-brown consolidated fluvial sediments rich in clay and silt can be observed at the base of the sections (silt member 2). These old deposits are overlain by greyish silty-clayey fine sands (silt member 1) which were deposited in a wide valley section and which are to be found up to 20 m above the recent river bed. A gravel pavement and/or remnants of slope debris covers the higher parts of the small interfluve ridges where the greyish deposits resisted gullying. Remnants of the brown silty and weakly cemented fine sands (SWD member 2) cover the slope debris remnants in some places. This is the case especially on the slightly inclined surfaces near the foot slope of rock walls where denudation and erosion processes are weak. A relatively dense vegetation cover prevents the removal of the SWDs.

The brown loose SWDs (SWD member 1; Fig. 5) that were transported and accumulated by the Hoanib flash floods into the gully relief, are to be found up to 5 m above the recent river channel. These silty fine sands (fine sand: ca. 64%; coarse silt: ca. 25%; clay: ca. 8%; colour: 7.5YR5/4) comprise 0.0% coarse sand (2 - 0.63 mm ∅) and less than 0.5% middle sand (0.63 - 0.2 mm ∅). At the base of these brown SWDs are thin layers of redeposited greyish silty sands, documenting an erosion phase in the gully relief, immediately preceeding the accumulation of the brown SWDs. The lowermost layers of the brown SWDs are interca-

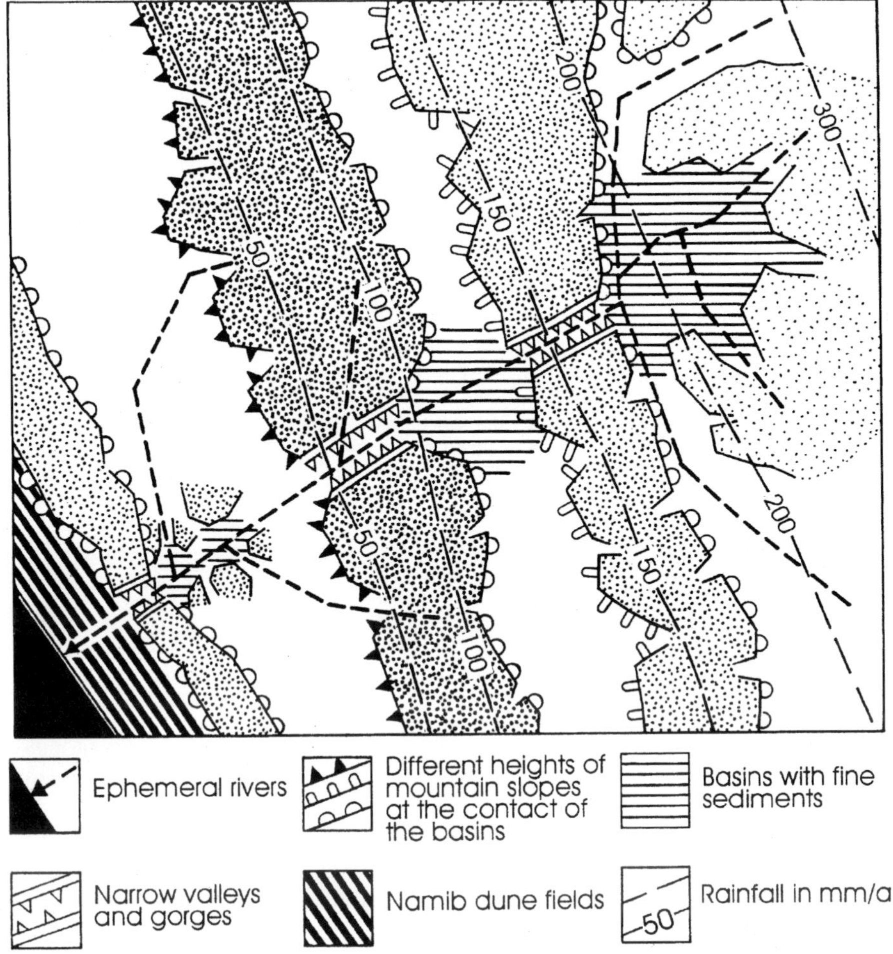

Fig. 4. Schematic morphological sketch of the valley structure of the Hoanib, Kaokoveld (after LESER 2000)

lated with redeposited greyish silty sands, thus giving evidence for intensive slope wash in the Khowarib Gorge at the time of the first accumulation of the brown SWDs. All sedimentary features associated with the brown SWDs prove that the slack water sedimentation occurred in a back-flooded embayment. A sedimentation as floodout sediments ('Flutauslaufsedimente'), in the sense of EITEL et al. (1998) and BLÜMEL et al. (2000a), can be excluded.

Two ^{14}C ages document a sedimentation during the last 500 years (Table 1, Fig. 5). As these young brown SWDs are partly eroded and are covered by a 5 to 10 cm thick layer of sheet wash material, we conclude that the accumulation of the SWDs occurred at least several decades to some centuries ago.

Table 1. ^{14}C ages related to slack water deposits (SWDs) of Namibian Desert valleys and to desert gypcrete sediment sequences

Hv	Sample	Material	Depth m	$\delta^{13}C$ ‰	^{14}C years BP	^{14}C content pmc	Calibration to calendar years; cal...	Sedimentary environment
23584	099-6	wood	0.5	-22.5	130 ± 80		AD 1665-1955	Khowarib Gorge, SWD
23585	099-7	wood	0.2	-23.8	210 ± 175		AD 1480-1955	Khowarib Gorge, SWD
23586	099-9	wood	3.0	-21.6	105 ± 85		AD 1675-1955	Khowarib Gorge, SWD
23587	099-10	wood	2.8	-24.4	410 ± 65		AD 1435-1620	Khowarib Gorge, SWD
23588	099-11	wood	1.8	-25.4	1810 ± 55		AD 135-320	Khowarib Gorge, SWD
23589	099-12	wood	0.8	-24.6	135 ± 55		AD 1675-1955	Khowarib Gorge, SWD
23590	099-13	wood	1.0	-22.1	120 ± 65		AD 1675-1955	Khowarib Gorge, SWD
23591	099-14	wood	1.0	-26.0	150 ± 50		AD 1670-1950	Khowarib Gorge, SWD
23592	099-16	charcoal	0.3	-25.2		106.7 ± 0.9	AD 1959/60	Khowarib Gorge, SWD
23593	099-39	wood	-	-19.0		131.8 ± 0.9	AD 1962 or 1978	Swakop/Khan, SWD
23594	099-40	wood	0.5	-23.2		127.6 ± 2.4	AD 1961 or 1980	Swakop/Khan, SWD
23595	099-42	wood	0.4	-21.9	95 ± 85		AD 1675-1955	Swakop/Khan, SWD
23596	099-43	wood	2.0	-24.1	130 ± 110		AD 1660-1955	Swakop/Khan, SWD
23597	099-44	wood	1.0	-23.5	100 ± 75	99.7 ± 0.9	AD 1675-1955	Swakop/Khan, SWD
23598	099-45	wood	0.2	-24.7		109.8 ± 0.8	AD 1960 or 1996	Swakop/Khan, SWD
23599	099-73	wood	0.6	-23.1		122.8 ± 1.1	AD 1961 or 1981	Kuiseb, SWD
23600	099-75	wood	0.5	-23.0		120.2 ± 0.9	AD 1961 or 1984	Kuiseb, SWD
23601	099-76	reed casts	1.0	- 6.3	4625 ± 95		BC 3505-3135	Kuiseb, SWD
23602	099-92	charcoal	1.0	-21.7	630 ± 165		AD 1255-1440	Tsauchab, SWD
23603	K00 587	wood	0.5	-21.0		101.0 ± 0.8	AD 1958	Sossus vlei, sand dune
23604	099-46	wood	0.4	-24.6		125.7 ± 1.0	AD 1963 or 1981	Swakop/Khan, SWD
23605	099-48	wood	0.5	-24.2		126.1 ± 0.9	AD 1963 or 1981	Swakop/Khan, SWD
21192	K00 905	wood	0.0	-23.3	150 ± 50		AD 1670-1950	Tsondab vlei, dead Acacia
22507	SWA 9-12	org. mat.	0.24	-11.9	675 ± 115		AD 1260-1405	
22508	SWA 9-11	wood	0.195	-14.0	500 ± 135		AD 1310-1615	
22509	SWA 9-8	org.mat.	0.18	-10.1	250 ± 160	96.9 ± 2.0	AD 1460-1955	Namib surface gypcretes
22510	SWA 9-1	org. mat.	0.32	-11.4	1385 ± 170		AD 540-850	
22511	KCA 10-1	org. mat.	0.12	-18.1		117.6 ± 1.7	AD 1962 or 1986	
22512	GOR 12-1	org. mat.	0.165	-21.9	115 ± 125	98.6 ± 1.6	AD 1660-1955	
22513	KOO 927	wood	2.0	-23.4	535 ± 60		AD 1395-1435	Amspoort Silts
22514	KOO 635	org. mat.	0.1	-19.7		133.6 ± 2.2	AD 1963 or 1977	Tinkas Cave sediment
15959	KOO 871	org. mat.	1.0	-11.3	690 ± 110	91.7 ± 1.3		Okavango, overbank silts
15961	KOO 876	org. mat.	0.2	-7.3	190 ± 85	97.6 ± 1.0		Okavango, overbank silts
15965	KOO 802	org. mat.	1.0	-16.5		106.0 ± 2.2	AD 1959/1960	Ekuma, overbank silts
15966	KOO 813	org. mat.	1.0		3390 ± 175	65.6 ± 1.5		Ekuma, overbank silts
Beta134364	KOO 713	wood	1.3	-25.0		109.1 ± 0.7	AD 1959/1960	Swakop, SWD
Pta 4517	KOO 721	wood	0.0	-24.7	120 ± 50		AD 1820-1930 AD 1700	Namib surface, driftwood
Erl-2607	099-90	plant	0.5	-29.2	294 ± 80	96.4 ± 0.9	AD 1439-1694 AD 1723-1815 AD 1919-1940	Tsauchab, SWD
Erl-2718	KOO 636	plant	0.15	- 24.9	504 ± 78	93.9 ± 0.9	AD 1298-1520 AD 1567-1628	Tinkas Cave deposits

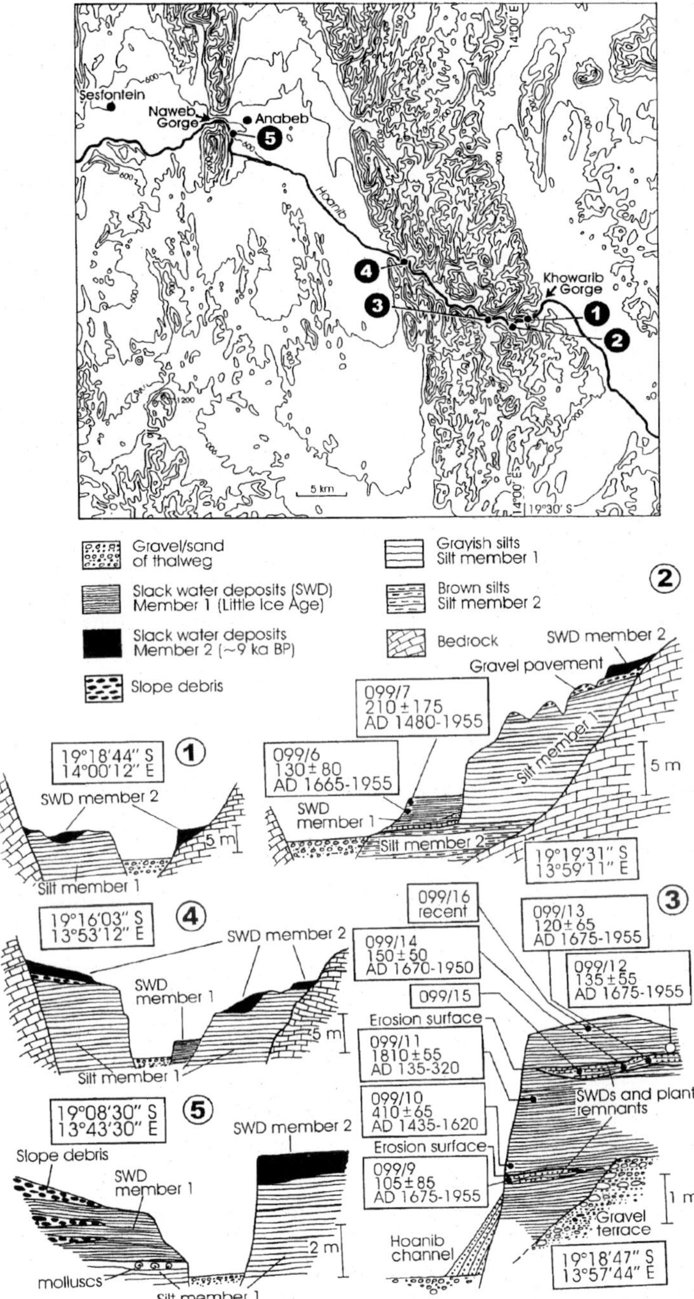

Fig. 5. The Hoanib in the Sesfontein area and Khowarib Gorge. Sections with SWDs; for details see text

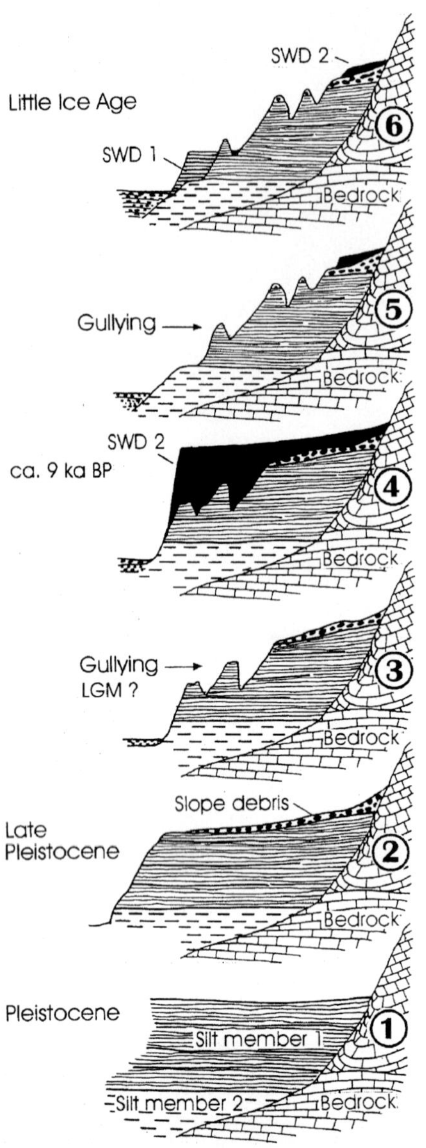

Fig. 6. Schematic development of the different SWD members; for details see text.

Downvalley at 19°18'47''S/ 13° 57'44''E, SWDs occur (Fig. 5) that are deposited in a widening of the valley and that are accumulated alongside the strongly eroded light greyish sediments. These greyish deposits outcrop the younger brown SWDs on the southern side of the valley. The actual thalweg is cut into the brown SWDs in a bend of the river channel. The outer bank consists of bedrock. Three sedimentation phases of SWDs are documented within the section by discontinuities in accumulation (erosion surfaces).The top of this sequence of SWDs again reaches about 5 m above the actual river channel. The sediments are brown in colour, laminated, unconsolidated and comprise predominantly silty-clayey fine sands (fine sand: ca. 40-50%; coarse silt: ca. 20%, clay: ca. 10%) with little $CaCO_3$ content and no components >2 mm ∅. Two horizons are rich in sticks, logs, seeds and other plant remains. Their sedimentation took place along the river's banks as the floods receded. This is a characteristic feature of SWDs. The sequence documents flash floods. ^{14}C dates of the organic rich layers show ages between AD 1675 and 1955. The two ^{14}C ages of samples of the central sediment unit document that organic material (logs) were redeposited by flash floods. Thus, the samples 099/10 and 099/11 cannot be used to determine the age of the slack water sedimentation. The sample 099/12 is driftwood; the ^{14}C age proves an accumulation phase between AD 1675 and 1955, too. A piece of charcoal from the uppermost (youngest) SWD has a recent age. It points to burning from the surface. The three SWD units are subdivided neither by fossil soils nor by marked erosion phases (gullying). The sedimentologic evidence as well as the ^{14}C dates document an accumulation during a relatively short phase.

The young brown SWDs of the two sites can be traced downvalley in the Khowarib Gorge. They often fill embayments and gullies which developed in the greyish sediments. This is also the case in the Hoanib valley around 19°16'03''S/13°53'12''E. The fact that in many places the young brown SWDs show intense erosion, is evidence that the age of the SWDs is at least one or several centuries.

A summary of the observations of the Khowarib Gorge reveal the following sequence of sedimentation and erosion phases from top to bottom (Fig. 6):

(6) Brown unconsolidated SWDs, accumulated ca. AD 1650 to 1955. SWD member 1.
(5) Phase of dissection, development of gullies.
(4) Deposition of brown silts, to date slightly consolidated, ca. 9 ka BP old. SWD member 2.
(3) Phase of dissection, development of gullies.
(2) Accumulation of slope debris.
(1) Deposition of greyish sands and silts of unknown age (Pleistocene), to date consolidated. Silt member 1. Underlain by red-brown consolidated silts of Pleistocene age. Silt member 2.

This sequence of sedimentation and erosion phases also developed near Anabeb (Fig. 5). East of Sesfontein, the Hoanib cuts through a mountain range and forms an antecedent river at the Naweb Gorge (BRUNOTTE & SPÖNEMANN 1997). Upvalley of the Naweb Gorge the Hoanib accumulated sediments rich in silt and clay which can be separated into greyish-white older silts (silt: ca. 60%; clay: ca. 30%; $CaCO_3$: ca. 30-35%) and brown younger silts (silt: ca. ca. 60-65%; clay: ca. 35%; $CaCO_3$: ca. 25-30%). The brown (7.5YR5/4) sediments can be correlated with the early Holocene deposits of the Khowarib Gorge (SWD member 2). Their thickness increases towards the Naweb Gorge. They only accumulated in front of the mountain range. An erosion phase is responsible for cutting the actual Hoanib channel into the greyish-white and the brown silts. Younger, unconsolidated SWDs were laid down up to 3 - 6 m above the actual Hoanib channel. They can be correlated with those SWDs of the Khowarib Gorge that were deposited during the last centuries (SWD member 1). In tributary mouths of the Hoanib valley coarse, angular gravels and boulders were deposited which intercalate with the younger SWDs. The gravels represent an intense erosion phase on the mountain slopes during the time of the SWD accumulation. There is no evidence that in recent times (during the last 100 or 200 years) debris material drained from tributary valleys into the main Hoanib channel.

In addition to the occurrence of these young SWDs there is evidence of slack water accumulation in adjacent areas. EITEL et al. (1999) describe a younger sedimentation phase in the basin west of Khowares (Khowarib Gorge) dated to 2.5+/-0.8 ka (TL date). RUST (1997) mentioned the Anabeb Silts of the same age (cal ^{14}C age: 382 B.C., Pta 7349) which spread from the Naweb Gorge into the basin of Sesfontein. It is not yet clear whether these silts (dated to ca. 2 - 2.5 ka) can be correlated with the youngest silt member (SWD member 1) of the Khowarib Gorge.

From the Hoanib valley near Amspoort VOGEL & RUST (1990) and RUST (1997) describe a 'forest' buried by fluvial silty sediments during the LIA. The sedimentation of the Amspoort Silts occurred since AD 1640 and continued probably till the 19th century, although a concentration of the accumulation to only 70 years (AD 1640 to 1710) cannot be excluded (VOGEL & RUST 1990). The dating of the sedimentation phase is based on numerous calibrated ^{14}C ages. The Amspoort Silts were deposited at the same time interval during which the SWD member 1 of the Khowarib Gorge accumulated. Therefore, we conclude that between AD 1640 and ca. 1800 as a consequence of frequent flash flood events the development of the SWDs in the Khowarib Gorge and in front of the Naweb Gorge occurred, as well as the deposition of the Amspoort Silts at the eastern edge of the Namib Desert (ca. 19°20' - 19°25'S / ca. 12°55' - 13°12'E).

5 The Swakop/Khan Rivers, central Namib Desert

In the area of the confluence of the Khan and Swakop Rivers, only the youngest SWDs were investigated. SWDs are observed as small terraces in places with little or no erosion during recent flash floods. The SWDs reach up to 5 m in the lower Khan Valley and up to 2 - 3 m in the lower Swakop Valley above the actual river channel. The bedded SWDs show in the upper part loose, unconsolidated younger deposits and in the lower part slightly consolidated older deposits (Fig. 7). Contrary to the situation in the Hoanib Valley, in the Swakop-Khan area no older Pleistocene, several metres thick, silty-clayey sediments have yet been found.

The two SWD units (terraces) can be traced over many kilometres in the area of the Swakop-Khan confluence. In the lower Swakop Valley, these SWD terraces disappear. According to the sedimentologic and pedologic observations they can be correlated to the SWD member 1 of the Hoanib Valley. The ^{14}C ages (Fig. 7) corroborate this, although a younger age con not be excluded.

Between the Atlantic Coast and the railroad bridge (ca. 6 km inland), many sedimentation and erosion phases alternated in the Swakop River valley during the last century (STENGEL 1964). In AD 1987, near the railroad bridge, a fluvial terrace ca. 2.5 m high existed, composed of alternating layers of silt and sand. This terrace is younger than AD 1960 (^{14}C age: < AD 1960, Beta 134 364) and presumably was formed during the rainy season of AD 1962/63. In 1999 this terrace was completely eroded by the Swakop River.

6 The Kuiseb River, central Namib Desert

HEINE et al. (2000) and HEINE & HEINE (2002) report on older SWDs in the Kuiseb Valley. The late Pleistocene Homeb Silts are subject of much debate about their paleoclimatic significance. Here, we focus on the youngest SWDs of the

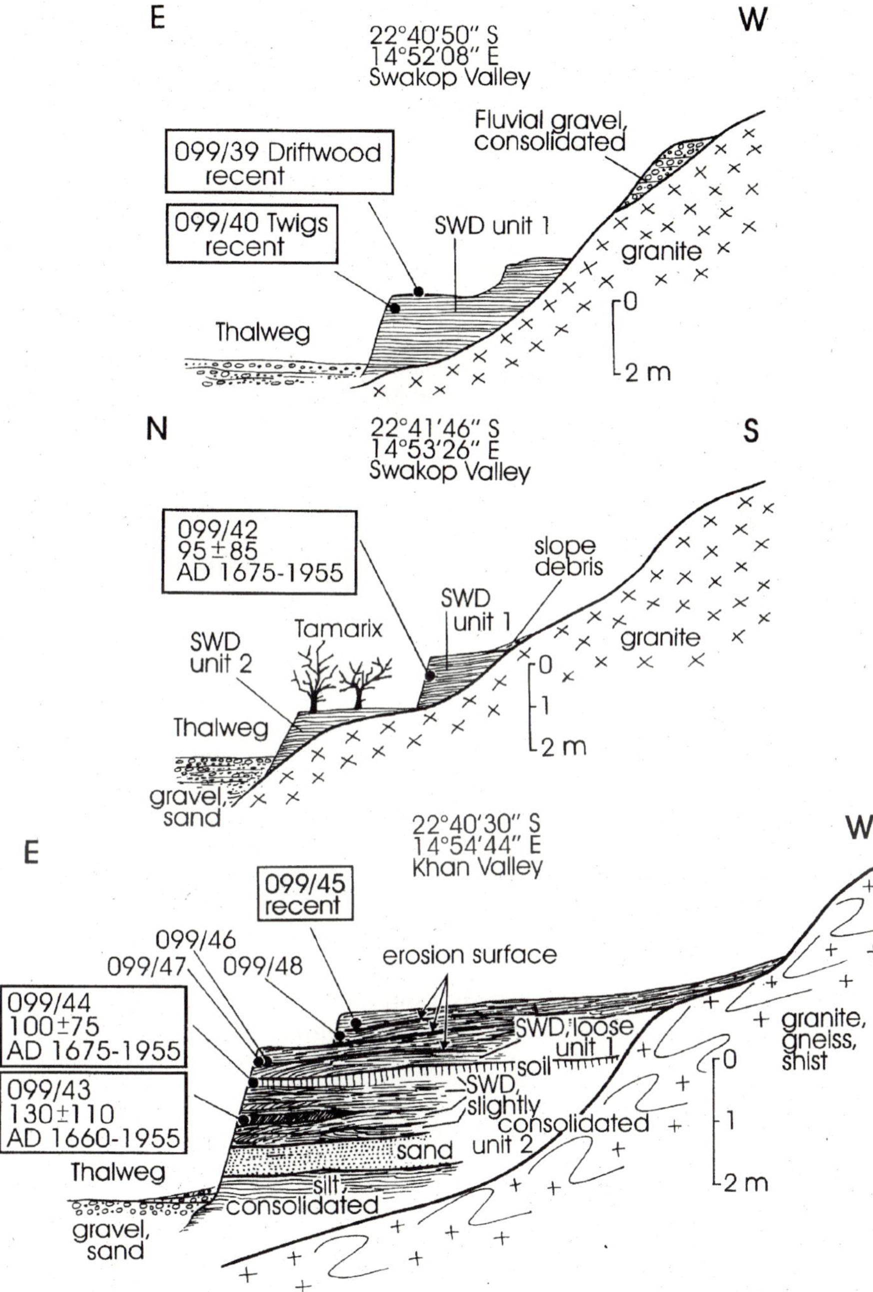

Fig. 7. SWD sections of the Khan and Swakop Valleys

Kuiseb. Between Homeb and the Gorob tributary mouth young unconsolidated SWDs accumulated. They form narrow terrace strips along the valley slopes in protected areas and rise 3 - 4 m above the river channel. On the basis of our field investigations and ^{14}C age determinations (Table 1), SWDs appear to have developed in the Kuiseb Valley only during most recent times (LIA and younger) and during the late glacial/early Holocene transition as well as during the LGM (VOGEL 1982; HEINE & HEINE 2002).

Along the lower Kuiseb River logs of driftwood occur high above the present river bed in the desert. The dating of these logs provides information on exeptionally large floods in the past (VOGEL & VISSER 1981). One log has a calibrated ^{14}C date between AD 1490 and 1630, two logs are younger than AD 1660, and one log has a conventional ^{14}C date of 940+/-35 yr BP (VOGEL & VISSER 1981).

Logs of dead wood are found on the Namib surface between Gobabeb and the Vogelfederberg. Flash floods (sheet flood processes) transported the logs. The calibrated ^{14}C date of a log is AD 1820 - AD 1930 or AD 1700 (Pta-4517) (HEINE & WALTER 1996).

7 The Tsauchab River, southern Namib Desert

Near Hauchabfontein in the Tsauchab Valley, SWDs accumulated and several SWD members can be distinguished. The young SWDs near Hauchabfontein form a small terrace of about 1.5 - 2 m height along the recent river channel. The terrace is cut into older late Pleistocene to early Holocene clayey, 10 to 15 m thick silts which fill a small basin. These older flood deposits are described by HEINE (1993).

The slightly consolidated late Holocene SWD member (fine sand: ca. 28%; silt: ca. 64%; clay: ca. 10%; $CaCO_3$: ca. 5-7%) shows several sedimentation cycles. Charcoal and small pieces of wood of the lower unit are dated to AD 1255-1440 (^{14}C, Hv 23 602) and AD 1439-1694/1713-1815/1919-1940 (AMS ^{14}C, Erl-2607). Sedimentation occurred during the LIA. The loose SWDs of the upper unit (fine sand: ca. 35%; silt: ca. 50%; clay: ca. 13%; $CaCO_3$: ca. 9%) were not dated. Compared to the sedimentologic characteristics of other SWDs in the Namib valleys, we conclude that they formed during most recent times.

8 Data from other sites of the Namib Desert

Deposits of flash floods of recent decades are to be found along modern river channels in many places in the Namib Desert and adjacent areas in the east (HEINE 1998a; HEINE et al. 2000; HEINE & HEINE 2002). In addition to data mentioned by HEINE (1998a) and HEINE et al. (2000) it is noteworthy that near the Palmwag Lodge in the Uniab catchment SWDs and fluvial gravels are intercalated. The SWDs (sand: 10%; silt: 70%; clay: 21%; $CaCO_3$: 0%) are free of $CaCO_3$ and thereby differ from all SWDs of other catchments. Only the SWDs of the Visrivier (Farm Nomtsas north of Maltahöhe) are also free of $CaCO_3$. The lack of carbonate depends on the rocks of the catchment area in which the SWDs occur. The SWDs near Palmwag originate in the $CaCO_3$ free Etendeka Basalts; the Nomtsas SWDs exclusively stem from carbonate-free shists and sandstones. The X-ray diffraction analyses of the Palmwag SWDs clearly document the dominance of the 14 Å minerals which were formed by weathering of basalts (Fig. 8). Thus the silts prove the close relationship between the petrography of the catchment (source area) on the one hand and the clay mineral association and carbonate content on the other hand. The observations of EITEL et al. (1998) and EITEL et al. (1999), who argue for a long-distance dust transport of the silty material of the river terraces, must be supplemented by these findings.

9 The gypcretes of the central Namib Desert

In the central Namib Desert the gypcretes show a regular composition independent of the relief position. All sections are layered profiles. The single units rarely exceed 10 cm in thickness. They mostly stem from para-autochthonal material. Most of the sediments were redeposited earlier than the main phase of calcrete development during Miocene (?) times. Some additional phases of intensive redeposition of surface material must have occurred during the Pleistocene. Hence, weathering debris was spread over the vast plains of the desert. These thin loose sediments form the material of the soil development in places where calcretes do not reach the surface (west of ca. 15°25'E). The soils are extremely weakly developed. Nevertheless, soil horizons are clearly visible that are due to pedogenic processes associated with the original lamination of the sediments. In the west, the surface sediments of the Namib Desert show a 1 - 2 cm thick redeposited layer, becoming thicker to the east. After precipitation events, this surficial layer may saturate with water and move in suspension from higher to lower ground. Beneath this layer gypsum accumulates and contributes to the formation of gypsum crusts (see HEINE & WALTER 1996). The investigation of the gypcrete sections demonstrate that after rains redeposition of the loose surface material and gyrcrete formation dominated the geomorphic and pedologic processes in the central Namib Desert, at least since late Pleistocene times. Gypcretes developed throughout the area of investigation (Fig. 9). Only the sediments in the major dry valleys show different sections.

In the east, calcretes are dominant. To the west, the calcretes convert to calcrete/gypcrete mixed crusts and, finally, to gypcretes underlain in some places by calcretes. The gypcretes are post-sedimentary formations cementing the redeposited material. Gypcretes are zonal desert soils which are developing further in recent times. A prerequisite is that sulphur-bearing atmospheric compounds, specifically dimethyl sulphide from the oceans, may be oxidized in the atmosphere and deposited on the desert surface (BAO et al. 2000).

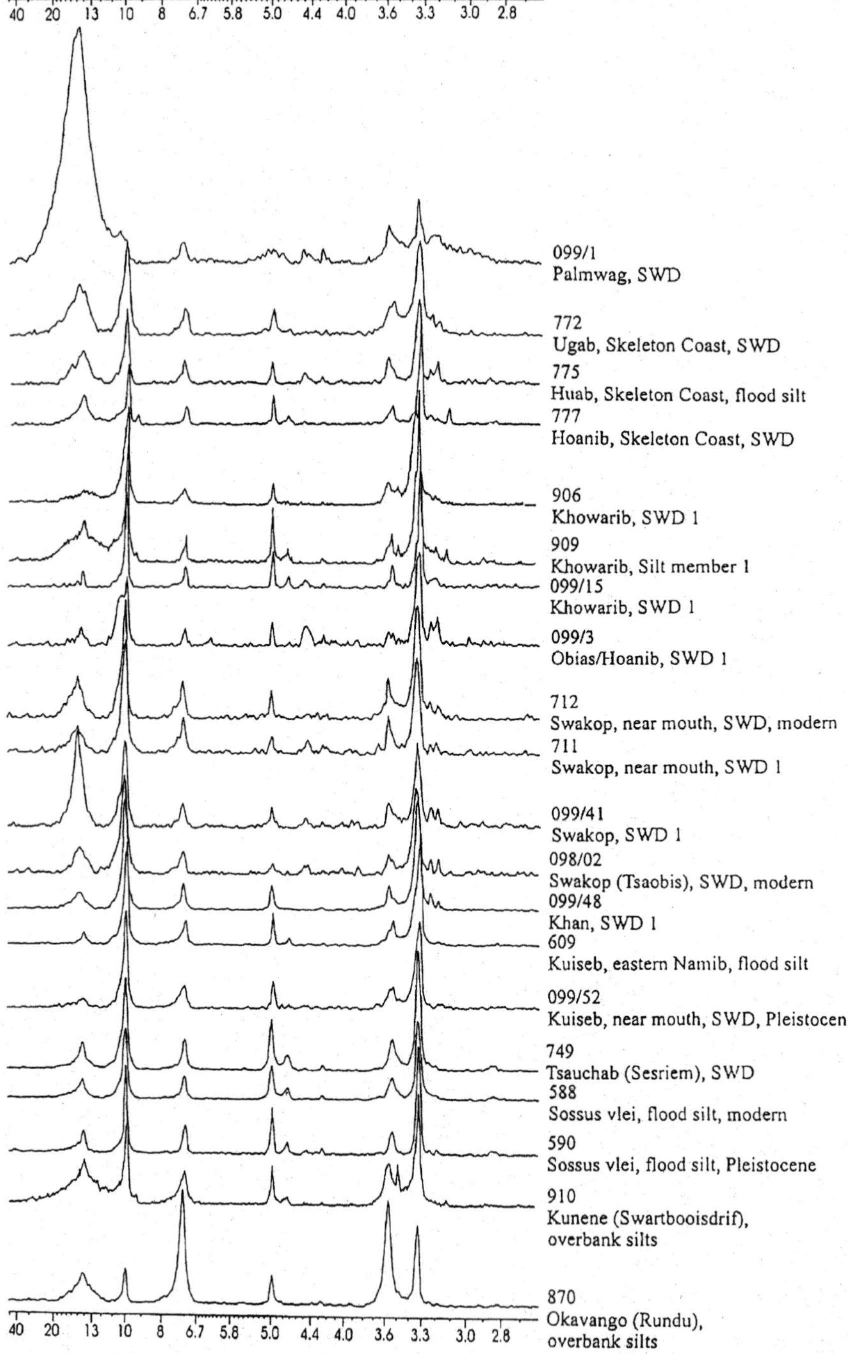

Fig. 8. X-ray diffractograms of different SWD and flood sediments. Clay samples (< 2μm)

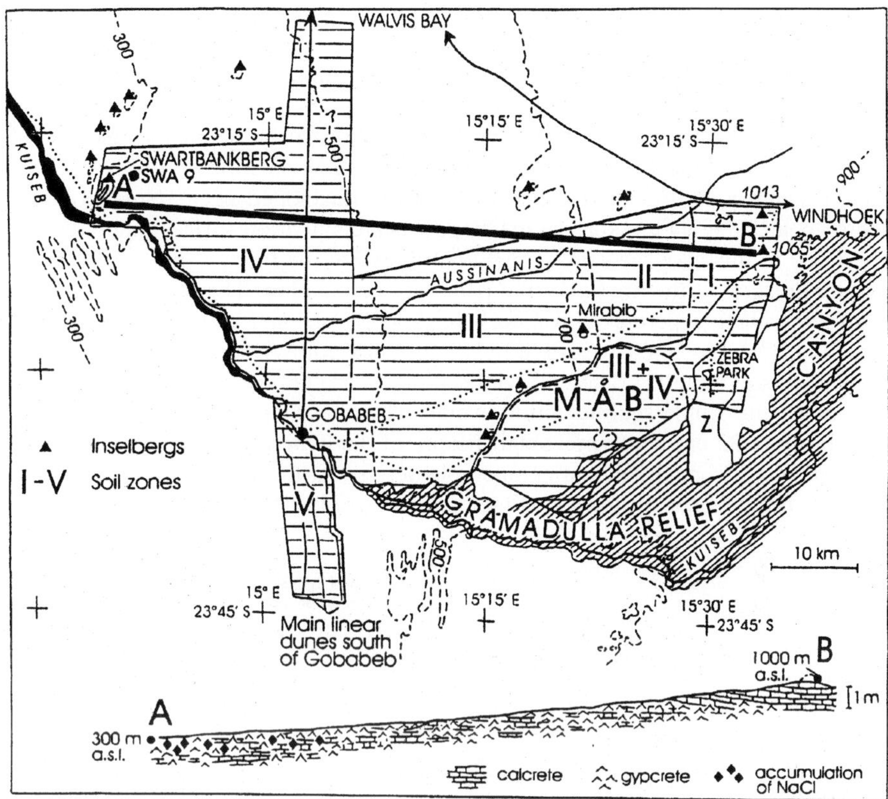

Fig. 9. Soil zones and zonal occurrence of duricrusts in the central Namib Desert (after R. WALTER, unpubl. data)

Investigations of the desert soils by R. WALTER (HEINE & WALTER 1996) document the zonal distribution of the different patterns of the gypcretes. The central Namib soils can be divided into five groups (zones): (1) Soil zone I characterises the calcrete areas in the east. (2) Soil zone II shows gypsum in the calcretes which are overlain by thick loose surface material (transition zone). (3) Soil zone III is characterised by regularly developed gypcretes beneath thin (<3 cm) layers of loose material. The amount of gypsum and pore volume increases to the west. Calcretes at the base of the sections are absorbed by the gypcrete development. Fissures in the gypcretes that do not contain gypsum and are filled with loose fine material, form polygonal structures. North of the Kuiseb valley several terraces or benches are formed that are cut into bedrock and lack predominant structural control. The gypcretes on these terraces show different stages of development related to the height above and the distance from the river channel. These observations point to little or no surface processes and thus to a long period of an extremely arid climate without interruptions by pluvial phases. (4) Soil zone III is composed of NaCl-rich carbonate/gypsum mixed crusts overlain by up to 30 cm thick loose material. (5) Soil zone V is characterised by eolian processes combined with pedogenic processes. This zone is more or less free of gypsum. The eolian morphodynamic processes hamper the accumulation of gypsum by blowing away initial gypsum aggregates.

The first ^{14}C dated gypcrete sections give information about the geomorphic processes during the LIA. North of the Swartbank Mountains on the Namib surface, a pan about 150 m long and 50 m wide, developed near a small ridge of schist rocks. The Namib surface developed as a denudation surface since the Tertiary, yet, in this case the schist ridge contributed to the formation of the deflation pan by preserving the area near the pan from denudation. The section of the pan shows a typical gypcrete with a well marked fissure and a redeposited horizon (Fig. 10). The gypcrete is overlain by 19 cm thick colluvial sediments, although it belongs to soil zone IV. The colluvial sediments are laminated and can be associated to several single sedimentation events. The oldest ^{14}C age refers to a decomposed root in 32 cm depth in a small fissure of the gypcrete. Two similar samples were gathered from fissures at comparable depths. At the time the plants grew the pan was not covered by loose sediments. Presumably plant growth was restricted

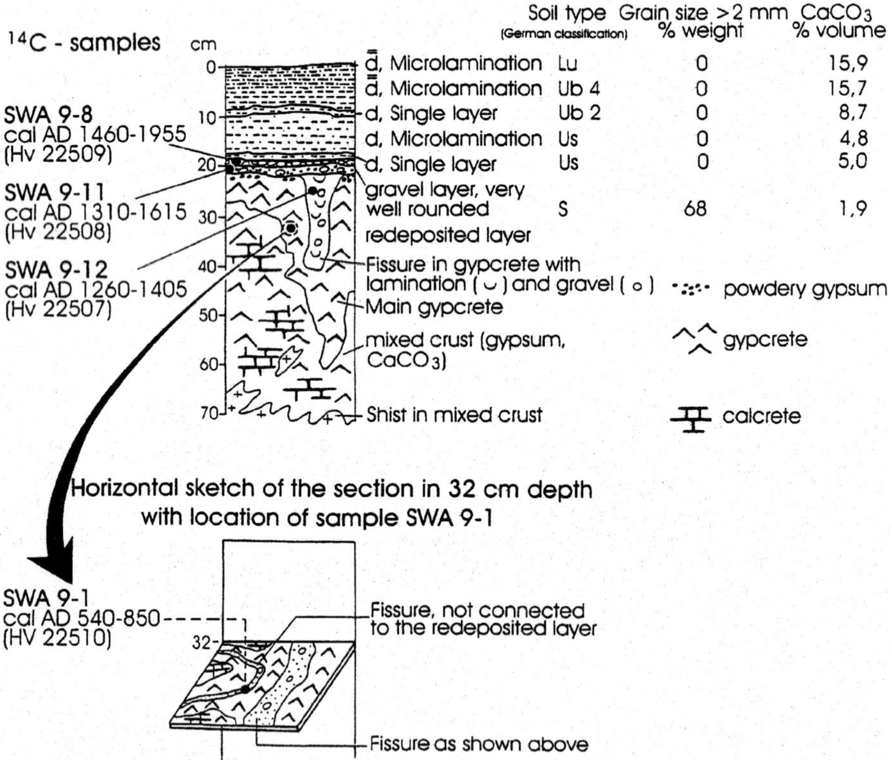

Fig. 10. Soil profile of the central Namib Desert in a small pan (after R. WALTER, unpubl. data)

to gypsum-free fissures filled with fine material. The fissure was closed after cal AD 540-850 (Hv 22 510). Then the bigger fissure opened. This succession of processes is documented in characteristic forms of recent fissures. The redeposited horizon at the top of the section is undisturbed.The minimal age of the development of the fissure gives the date cal AD 1260-1405 (Hv 22 507). The date cal AD 1310-1615 (Hv 22 508) suggests the minimal age of the beginning of the colluvial sedimentation in the pan documented by the gravel layer. The gravel reaches the surface near the edge of the pan. In the centre of the pan the gravel layer is overlain by fine sands and silts. The age cal AD 1460-1955 (Hv 22 509) is a maximal age for the accumulation of the laminated sheet flood sediments, because the dated material is cut by the overlying sediments.

The interpretation of the pan section with respect to our knowledge of the Namib soils and relief forms (HEINE & WALTER 1996) suggests that during the LIA the morphologic processes were quite different from those of the earlier Holocene. The pan has no outlet and was deflated until the beginning of the LIA. Several flash flood events contributed to the filling of the pan with the basal coarse sediments and the upper fine sands and silts. These laminated sediments represent precipitation events of a high magnitude, which caused the redeposition of the loose surface material on the Namib surface and into the pan. Since the relatively heavy rains of 1997 in the Namib Desert did not contribute to the redeposition of surface material near the pan, although water filled the pan, the laminated sediments point to even more extreme precipitation events.

These observations are corroborated by a date of another section. In a little valley (3rd order, 23°32,27'S, 15°23,63'E), organic material accumulated in 16.5 cm depth beneath well-sorted fine gravelly sediments. Similar sediments are observed in many comparable channels. The age of cal AD 1660-1955 (Hv 22512) points to redeposition phases during the LIA.

Our investigation in the central Namib Desert prove that, during the Holocene, a marked change of the geomorphodynamic processes occurred only during the LIA. The gypcrete sections show no disturbances in gypcrete formation. This is interpreted as evidence for the long late Pleistocene extreme aridity in the central Namib Desert without any break by pluvial episodes. Neither the Younger Dryas oscillation nor late Quaternary/early Holocene precipitation fluctuations can be observed. Our results prove for the first time that the LIA caused changes in the geomorphodynamic processes in the central Namib Desert. The climate of the LIA is characterised by repeated extreme flash floods in the desert itself. Due to the characteristics of the soil sections we cannot determine whether there were higher amounts of annual precipitation rates.

A change of the ground water situation in the Namib Desert over about 500 years is reflected in various records. The fact that the *Welwitschia mirabilis*, a groundwater dependent plant, show no baby plants in the central Namib Desert points to a lowering of the groundwater table and/or a change in the annual distribution (variability) of the rains. In the Sossus Vlei and Tsondab Vlei areas, dead *Acacia erioloba* trees also point to a lowering of the groundwater table over the last 500 years (VOGEL 1987; HEINE 1995; and ^{14}C age of dead wood of a Tsondab Acacia: 150+/-50 BP, Hv 21 192). In the Tinkas Cave (22°48'25''S,

15°03'55''E, ca. 425 m a.s.l.), plant material was removed from blown-in sandy silts. The AMS ^{14}C age of a wood fragment is 504±78 BP (sample KOO 636: ERL-2718, Table 1). The dated sandy silts represent the last phase during which sand, silt and plant remnants were transported by water into the cave. Only sheet floods could have caused these processes.

10 Discussion and conclusion

Flash floods cause the accumulation of SWDs with characteristic features as described by HEINE (1998a), HEINE et al. (2000) and HEINE & HEINE (2002). The documentation by JACOBSON et al. (1995) confirms this interpretation, especially from the numerous photos of the AD 1995 flash floods in the northern Namib Desert. The observations of flash floods during the last decades show that organic material was carried downstream in flood waters together with sands and silts. Similarly, former flash floods deposited the sediments found in the silt terraces of the Namib valleys. Cross-sections of old floodplains reveal alternating layers of sands, silts and organic matter. Only flash floods contribute to the accumulation of fluvial silts away from the - mostly - sandy-gravelly river channel. The areas of the riparian vegetation along the Namib river channels show up to 50 cm thick layers of SWDs after one big flood. The same floods that deposit the SWDs next to the main channel flow in overbank areas may cause erosion in certain reaches of the river channels. The so-called silt terraces of the Namib valleys are neither river endpoint accumulations ('Flutauslaufsedimente') nor do they stand for greater aridity in the upper reaches of the catchment as is thought by many authors (e.g. VOGEL 1989; BLÜMEL et al. 2000a; EITEL et al. 1998, 1999; RUST 1997). The majority of the silt terraces are SWDs in the sense of KOCHEL & BAKER (1982) and ZAWADA (1997) and, therefore, document big floods.

The SWDs are records of extraordinary discharge events of the Namib rivers. The SWDs mainly accumulated during the LIA. In the Hoanib catchment, the SWDs indicate repeatedly big flash floods between AD 1650 and the 19th century. These flash floods originate because of extremely heavy rainfall in the upper catchment. The intercalation of slope debris with SWDs show that at the time of the slack water accumulation in the Khowarib Gorge and the Sesfontein area, extreme precipitation events occurred. Near Amspoort, where the Hoanib leaves the narrow epigenetic valley, the Amspoort Silts (VOGEL & RUST 1990) were deposited as a consequence of the valley widening. Here, the cross-section of the channel becomes wider, the capacity for carrying sediments is reduced (see SCHEIDEGGER 1990; HEINE et al. 2000). After accumulating the silts, the waters discharged further west across the Namib desert and might have reached the Atlantic Ocean. Therefore, the Amspoort Silts represent great floods and the sediments can be described as 'floodouts'. A floodout can be defined as 'a site where channelized flow ceases and floodwaters spill across adjacent alluvial surfaces' (TOOTH 1999). Like the SWDs of the Khowarib Gorge, the Amspoort Silts (as floodouts) document flash flood events during the LIA. The Amspoort Silts were deposited within several decades (or a few centuries at least). The volume of the Amspoort Silts in relation to the time of accumulation shows that only flash floods of the Hoanib river could have caused the sedimentation of these stacked silt units. During recent flash floods the thickness of the single SWD layers is thinner than the thickness of most single layers of the Amspoort Silts. Therefore, we do not agree with VOGEL & RUST (1990) and RUST (1997) that the Amspoort Silts represent river end point sediments ('Flutauslaufsedimente') that depend on more arid climatic conditions (less precipitation) in the upper reaches of the catchment. Their findings are not corroborated neither by the hydraulic geometry of river channels and the dynamics of flowing water (SCHEIDEGGER 1990; HEINE et al. 2000) nor by observations of flash flood sedimentation during recent flood events (JACOBSON et al. 1995).

The erosion of the Hoanib channel cut into the Amspoort Silts is interpreted as a result of gullying by slackening waters at the end of the flood discharge (see HEINE & HEINE 2002) as well as by minor flood events. Thus the erosion channels that are cut into the Amspoort Silts are not interpreted as evidence for more humid phases in the catchment, as VOGEL & RUST (1990) and RUST (1997) assume, but rather as records of gullying by slackening waters and minor floods.

The SWDs as well as the alluvial sediments of the Namibian valleys show great differences in their clay mineral associations, which represent the great variety of bedrock of the different catchments (Fig. 8). The material of the SWDs stems from weathering residuals of the interfluve areas upstream and is washed by sheet floods into the river channels. In the main channels the flash floods carry the fine silty suspended sediments in turbulent flow. Where the flow velocity decreases, laminar flow occurs. This can be observed in embayments, behind obstacles (rock outcrops), in valley widenings, and at tributary mouths etc. Sedimentation of SWDs is the result. The X-ray diffractograms show no difference in clay mineral associations of certain valleys with respect to the age of the SWDs. Hence, we assume that the conditions for the development, transport and deposition of the SWDs did not change since the late Pleistocene.

Since the termination of the last glacial, the accumulation of flash flood sediments in the valleys of Namibia occurred (1) during the early Holocene (ca. 8-10 ^{14}C ka BP; HEINE 1995, 1998a) and (2) during the Neoglacial, especially during the LIA. It is obvious that until now no SWDs are known that accumulated between ca. 8 ^{14}C ka BP and the onset of the Neoglacial.

The SWDs of Neoglacial and LIA times seem to have been developed in greater thickness and wider distribution in the valleys of the northern Namib (Hoanib, Hoarusib, see VOGEL & RUST 1987; RUST 1997) than in the valleys of the central Namib (Swakop, Khan, Kuiseb). This points to the fact that the extreme precipitation events occurred above all in the northwestern areas of Namibia. The weather conditions may have been similar to those of the year 1995 when big floods occurred (JACOBSON et al. 1995). Presumably, the rains were heavier during the early Holocene and the Neoglacial.

A careful evaluation of the paleoclimatic evidences from Namibia (HEINE 1995; HEINE 1998a; HEINE et al. 2000) do not reveal pronounced climatic and/or hydrologic changes during the Holocene (see also SCOTT et al. 1991; HEINE 1995). Only during the early Holocene (ca. 10 to 8 ^{14}C ka BP), were the eolian processes more active with a remarkable short windy (dune forming) episode about 8.3+/-0.8 ka BP (TL age HEINE 1995). Further evidence for this short arid period comes from the observation of the sedimentation of loess-like material in the Kaokoveld (BRUNOTTE & SANDER 2000a; TL age ca. 8 ka) and of the mobilisation of dune sand in the southwestern Kalahari where dune building processes stopped after ca. 8 ka BP (EITEL et al. 2000; TL age: 9.5+/-1.8 ka, 8.8+/-1.2 ka and 8.0+/-0.6 ka). According to our investigations the period of accelerated eolian activity matches a period characterised by silt accumulation, e.g. a period with frequent flash floods and slack water deposition. Flood deposits developed in the *vleis* (fertile wetland) of the Sossus and Tsondab areas which are dated to 10-8 ^{14}C ka BP. The Natab Silts of the Kuiseb valley are dated to 8-10 ^{14}C ka BP (VOGEL 1982, 1989; HEINE et al. 2000). In the northern Namib valleys (Unjab, Hoanib, VOGEL & RUST 1987), no SWDs accumulated during the early Holocene (RUST & VOGEL 1988). These observations reflect different precipitation fluctuations in space and time. During the period with stronger eolian activity in the western Kalahari, at about 10 - 8 ^{14}C ka BP, in the central Namib (Kuiseb, Tsondab, Tsauchab) more frequent flash floods occurred. At the same time, the northern Namib did not experience more frequent flood events. On the other hand, during the LIA the northern Namib especially had numerous big floods, whereas the central Namib was not affected by such big floods. All known and dated SWDs demonstrate that during the Neoglacial more frequent extreme flood events occurred. The sheetwash sediments of the Namib Desert surface and the gypcrete development give further evidence for the Neoglacial/LIA flash floods. Neoglacial silt accumulation as a result of floods is reported from the lower Kunene valley (silt sedimentation between cal AD 650 and 1955: BRUNOTTE & SANDER 2000b). Flood deposits of the Okavango between Rundu and the Popa Falls have conventional ^{14}C ages between 690+/-110 (Hv 15 959) and 190+/-85 (Hv 15 961) ka BP (HEINE, unpubl. data). In the Ekuma channel (Etosha Pan area), with the beginning of the Neoglacial, fluvial sand was deposited for the first time during the Holocene (3390+/-175 ^{14}C yr BP, Hv 15 966). This suggests that bigger floods occurred in the Ekuma catchment.

11 Paleoclimatic implications

The paleoclimatic interpretation of the SWDs and the Namib gypcretes suggests the following paleoclimatic development. During the early Holocene eolian processes were active in the west Kalahari. At the same time SWDs, caused by flash floods, accumulated in the central Namib Desert. Between ca. 8 and 3.5 ^{14}C ka BP there is no record of major flood events nor of eolian sand mobilisation. PARTRIDGE et al. (1999) reported that during the Altithermal (8-6 ka BP) the precipitation was reduced by 10 - 20% in the Namibian highland and the eastern Kalahari. If this is true, then this climatic fluctuation was not characterised by flash flood events. During the Neoglacial, especially during the LIA, the flood events increased in number and intensity in the northern Namib Desert, whereas in the central Namib Desert flood events played a minor part. An increase in eolian activity occurred in Neoglacial times. Presumably, this was caused by climatic and/or anthropogenic factors.

If the Amspoort Silts represent the highest rainfall during the LIA, then this interval with extreme flash floods broadly coincides with the lowest LIA temperatures in mid-latitude regions of the northern hemisphere. This implies that the latitudinal pattern of century-scale climate anomalies during the past 1000 years was opposite to that which occurred on millennial timescales during the last glaciation (HEINE 1983; VERSCHUREN et al. 2000). The highest inferred rainfall events (flash floods) of the past 1000 years coincided with the Maunder minimum of solar radiation (AD 1645 - 1715). In continental East Africa, the long-term history of water-resource availability reflects an interval with above-average rainfall around AD 1700 (VERSCHUREN et al. 2000). In the South African summer rainfall region, geomorphological interpretations relating to pans, springs and vleis, as well as palynological and micromammalian evidence, indicate trends towards wetter conditions during the LIA associated with negative sea-surface anomalies (COHEN & TYSON 1995: 310). Annual layer thickness of stalagmites from Drotzky's Cave, Botswana, show marked variations of rainfall and soil moisture conditions in the summer rainfall area of the Kalahari (BROOK 1999).

We suggest that the precipitation anomalies (flash floods) of the Neoglacial and the LIA, respectively, are associated with circulation anomalies over southern Africa. We conclude that a significant proportion of summer rainfall is derived from tropical-temperate troughs (TTTs, TODD & WASHINGTON 1999) which extend over both continental southern Africa and the adjacent southwestern Indian Ocean. By using daily data TODD & WASHINGTON (1999) found that rainfall associated with TTTs over southern Africa results from distinct patterns of anomalous low-level moisture transport, which extends to the planetary scale, notably across the equatorial Indian Ocean. The principle mode of precipitation variability is a dipole structure with bands of rainfall orientated northwest to southeast across the region. The position of the temperate trough and the TTT cloud band alternates between the southwestern Indian Ocean and the southeast Atlantic. The synoptic scale TTT events over southern Africa/southwestern Indian Ocean often result from large-scale planetary circulation patterns. Tropical and extratropical dynamics are involved in producing these TTT cloud bands over southern Africa, influence their regional occurrence and, hence, the distribution of rainfall in the Namib Desert.

Tropical-temperate trough systems are the dominant rain-producing synoptic type over southern Africa (TODD & WASHINGTON 1999). We conclude that the LIA climate fluctuations are bound to regional and global atmospheric circulation and moisture flux patterns. For example, warm-season temperatures in Tasmania are related to large scale sea surface temperatures in the Indian Ocean and eastwards to the dateline (COOK et al. 2000). The record from the Namib Desert, indicating that the SWDs from the northern Namib reflect more frequent and more intense floods during the LIA than in the central Namib Desert, proves the interrelation between the extreme flood events and the TTTs. Winter rains (see WAIBEL 1922) do not seem to have influenced the flash floods and the slack water deposition. During the LIA, TTTs affected the northwestern Namib Desert more frequently than during the Altithermal and the last glacial maximum (LGM) Our knowledge of the LIA circulation pattern over southern Africa and the adjacent oceans is not yet clear. The situation is quite complex. That is shown by the results of ENSO events. These occurred in the relatively cold background of the pre-1970 period in the southern oceans, and they had only little effect on the rainfall conditions in southern Africa. In contrast, more recent ENSO events, with warmer SSTs (sea surface temperatures) over the southern oceans, led to a climatic bipolar pattern between continental southern Africa and the western Indian Ocean (RICHARD et al. 2000).

Acknowledgements

I am grateful to the Deutsche Forschungsgemeinschaft for financial support for research in southern Africa (DFG, He722/14-6). Many people helped with field work and laboratory analyses. I am indebted to the government of Namibia (Ministry of Environment and Tourism, H. Kolberg) for research permits and to the DRFN (Desert Research Foundation of Namibia, J. Henschel) for their cooperation. Special thanks to M.A. Geyh (Hannover) and W. Kretschmer (Erlangen) for age determinations and discussions, to R. Walter (München) for extensive research on the Namib gypcretes, to J. Völkel (Regensburg) for X-ray diffractograms, to T. Kühn (Berlin), C. Heine (Eberswalde) and A. Heine (Regensburg) for assistance in the field, to A. Reuther and P. Chifflard for soil and sediment analyses.

References

Bao, H., Thiemens, M.H., Farquhar, J., Campbell, D.A., Chi-Woo Lee, C., Heine, K. & Loope, D.B. (2000): Anomalous ^{17}O compositions in massive sulphate deposits on the Earth. - Nature 406: 176-178.

Blümel, W.D., Hüser, K. & Eitel, B. (2000a): Landschaftsveränderungen in der Namib. - Geographische Rundschau 52(9): 17-23.

Blümel, W.D., Hüser, K. & Eitel, B. (2000b): Uniab-Schwemmfächer und Skelettküsten-Erg: Zusammenspiel von äolischer und fluvialer Dynamik in der nördlichen Namib. - Regensburger geogr. Schr. 33: 37-55.

Bradley, R.S. (2000): Climate Paradigms for the Last Millennium. - PAGES Newsletter, vol. 8(1): 2-3.

Brook, G.A. (1999): Arid Zone Palaeoenvironmental Records from Cave Speleothems. - In: A.K. Singhvi & E. Derbyshire (eds.), Paleoenvironmental Reconstruction in Arid Lands. Balkema - Rotterdam/Brookfield, 217-262.

Brunotte, E. & Sander, H. (2000a): Loess accumulation and soil formation in Kaokoland (Northern Namibia) as indicators of Quaternary climatic change. - Global and Planetary Change 26: 67-75.

Brunnote, E. & Sander, H. (2000b): Das Alter der Auensedimente des Kunene zwischen Epupa und Ruacana (Namibia): erste ^{14}C-Daten. - Zbl. Geol. Paläont. Teil I (1999), H. 5/6: 615-621.

Brunotte, E. & Spönemann, J. (1997): Die kontinentale Randabdachung Nordwestnamibias: eine morphotektonische Untersuchung. - Petermanns Geogr. Mitt. 141: 3-15.

Buch, M.W., Rose, D. & Zöller, L. (1992): A Tl-calibrated pedostratigraphy of the western lunette dunes of Etosha Pan/Northern Namibia. - Palaeoecology of Africa 23: 129-147.

Chambers, F.M., Ogle, M.I. & Blackford, J.J. (1999): Palaeoenvironmental evidence for solar forcing of Holocene climate: linkages to solar science. - Progr. Phys. Geogr. 23(2): 181-204.

Cohen, A.L. & Tyson, P.D. (1995): Sea-surface temperature fluctuations during the Holocene off the south coast of Africa: implications for terrestrial climate and rainfall. - The Holocene 5(3): 304-312.

Cook, E.R., Buckley, B.M., D'Arrigo, R.D. & Peterson, M.J. (2000): Warm-season temperatures since 1600 BC reconsructed from Tasmanian tree rings and their relationship to large-scale sea surface temperature anomalies. - Climate Dynamics 16: 79-91.

Eitel, B., Hüser, K. & Blümel, W.D. (1998): Feinsedimentterrassen am Ostrand der Namib: Entstehung und paläoklimatische Interpretation. - Erlanger geol. Abh., Sonderband 2 (Sediment '98): 19-20.

Eitel, B., Blümel, W.D., Hüser, K. & Zöller, L. (1999): Dust and river silt terraces at the eastern margin of the Skeleton Coast Desert (northwestern Namibia): genetic relations and palaeoclimatic evidence. - In: E Derbyshire (ed.), Loessfest '99, Bonn - Heidelberg 25 March - 1 April 1999, Royal Holloway, Centre for Quarternary Research, Extended Abstracts, p. 56-59.

Eitel, B., Blümel, W.D. & Hüser, K. (2000): The Early to Mid-Holocene Climatic Change in Southwestern Africa: Evidence from dunes, loess-like deposits and soils based on TL and OSL-data. - In: F. Diaz del Olmo, D. Faust & A.I. Porras (eds.), Environmental Changes During the Holocene, INQUA-Meeting 27-31 March 2000, Seville (Spain): 41-43.

Grove, J.M. (1988): The Little Ice Age. - Methuen - London, 498 pp.

Heine, K. (1983): Spät- und postglaziale Gletscherschwankungen in Mexiko: Befunde und paläoklimatische Deutung. - In: H. Schroeder-Lanz (ed.), Late- and Postglacial Oscillations of Glaciers: Glacial and Periglacial Forms. Balkema - Rotterdam, 291-304.

Heine, K. (1993): Zum Alter jungquartärer Feuchtphasen im ariden und semiariden südwestlichen Afrika. - Würzburger geogr. Arb. 87: 149-162.

Heine, K. (1995): Paläoklimatische Informationen aus südwestafrikanischen Böden und Oberflächenformen: Methodische Überlegungen. - Geomethodica 20: 27-74.

Heine, K. (1998a): Climatic change over the past 135,000 years in the Namib Desert (Namibia) derived from proxy data. - Palaeoecology of Africa 25: 171-198.

Heine, K. (1998b): Late Quaternary climate changes in the Central Namib Desert, Namibia. - In: A.S. Alsharhan, K.W. Glennie, G.L. Whittle and C.G.St.C. Kendall (eds.), Quaternary Deserts and Climatic Change. Balkema - Rotterdam, 293-304.

Heine, K., Heine, C. & Kühn, T. (2000): Slackwater Deposits der Namib-Wüste (Nambia) und ihr paläoklimatischer Aussagewert. - Zbl. Geol. Paläont. Teil I (1999), H. 5/6): 587-613, Stuttgart.

Heine, K. & Heine, J.T. (2002): A paleohydrologic re-interpretation of the Homeb Silts, Kuiseb River, central Namib Desert (Namibia) and paleoclimatic implications. – Catena, 48:107-130.

Heine, K. & Walter, R. (1996): Die Gipskrustenböden der zentralen Namib (Namibia) und ihr paläoklimatischer Aussagewert. - Petermanns geogr. Mitt. 140: 237-253.

Hupfer, P., Grassl, H. & Lozan, J.L. (1998): Überblick: Warnsignale aus der Klimaforschung. - In: Lozan, J.L., Grassl, H. & Hupfer, P. (eds.), Das Klima des 21. Jahrhunderts, Wissenschaftliche Auswertungen, Hamburg, 419-426.

Jacobson, P.J., Jacobson, K.M. & Seely, M.K. (1995): Ephemeral Rivers and their Catchments: Sustaining People and Development in Western Namibia. - Desert Research Foundation of Namibia (DRFN), Windhoek, 160pp.

Kinghorn, S. (2000): Regen in Swakopmund 1934 - 2000. - Nachrichten - Museum/Sam Cohen Bibliothek Swakopmund. Gesellschaft für Wissenschaftliche Entwicklung, Jg. 32(2): 21-27, Swakopmund.

Kochel, R.C. & Baker, V.R. (1982): Paleoflood hydrology. - Science 215: 353-361.

Lamb, H.H. (1977): Climate, Present, Past and Future. Vol. 2 - Climate History and the Future. - Methuen - London, 835 pp.

Lancaster, N. (1989): The Namib Sand Sea. Dune forms, processes and sediments. - Balkema - Rotterdam, 180pp.

Leser, H. (2000): Methodische Probleme sedimentologischer Untersuchungen pleistozäner Sedimente im Kaokoveld (Namibia). - Regensburger geogr. Schr. 33: 19-36.

Lindesay, J.A. & Tyson, P.D. (1990): Climate and Near-surface Airflow Over the Central Namib. - In: M.K. Seely (ed.), Namib ecology: 25 years of Namib research, Transvaal Museum Monograph No. 7: 27-37, Pretoria.

Partridge, T.C., Scott, L. & Hamilton, J.E. (1999): Synthetic reconstructions of southern African environments during the Last Glacial Maximum (21 - 18 kyr) and the Holocene Altithermal (8 - 6 kyr). - Quaternary International 57/58: 207-214.

Pfister, C. (1999): Wetternachhersage. 500 Jahre Klimavariationen und Naturkatastrophen 1496-1995. - Haupt - Bern/Stuttgart/Wien.

Richard, Y., Trzaska, S., Roucou, P & Rouault, M. (2000): Modification of the southern African rainfall variability/ENSO relationship since the late 1960s. - Climate Dynamics 16: 883-895.

Rogers, J. (2000): Quaternary Evolution of the Benguela Coastal Upwelling System. - SASQUA Newsletter 29 (May 2000): 5.

Rust, U. (1997): Kurzbericht über subrezent-rezente Terrassen an Hoanib und Hoarusib (NW-Namibia). - 1. Workshop über "Probleme der Landschaftsgeschichte und Geoökologie Namibias", IUNFG, Universität Bayreuth 07.11.1997, Kurzfassungen u. Vorträge.

Rust, U. & Vogel, J.C. (1988): Late Quaternary environmental changes in the northern Namib Desert as evidenced by fluvial landforms. - Palaeoecology of Africa 19: 127-137.

Scheidegger, A.E. (1990): Theoretical Geomorphology. 3rd ed. - Springer, Berlin etc., 434pp.

Scott, L., Cooremans, B., de Wet, J.S. & Vogel, J.C. (1991): Holocene environmental changes in Namibia inferred from pollen analysis of swamp and lake deposits. - The Holocene 1(1): 8-13.

Stengel, H.W. (1964): Die Riviere der Namib und ihr Zulauf zum Atlantik. I. Teil: Kuiseb und Swakop. - Scientific Papers of the Namib Research Station, No. 22, Pretoria, 50pp.

Todd, M. & Washington, R. (1999): Circulation anomalies associated with tropical-temperate troughs in southern Africa and the south west Indian Ocean. - Climate Dynamics 15: 937-951.

Tooth, S. (1999): Floodouts in Central Australia. - In: A.J. Miller & A. Gupta (eds.), Varieties in Fluvial Forms, Wiley - Chichester, 219-247.

Tyson, P.D. (1986): Climatic change and variability in southern Africa. - Oxford University Press, Cape Town, 220pp.

Verschuren, D., Laird, K.R. & Cumming, B.F. (2000): Rainfall and drought in equatorial east Africa during the past 1,100 years. - Nature 403: 410-414.

Vogel, J.C. (1982): The age of the Kuiseb river silt terrace at Homeb. - Palaeoecology of Africa 15: 201-209.

Vogel, J.C. (1987): Chronological Framework for Palaeoclimatic Events in the Namib. - NPRL Research Report CFIS 145, Pretoria, 20pp.

Vogel, J.C. (1989): Evidence of past climatic change in the Namib Desert. - Palaeogeography, Palaeoclimatology, Palaeoecology 70: 355-364.

Vogel, J.C. & Rust, U. (1987): Environmental changes in the Kaokoland Namib Desert during the present millennium. - Madoqua 15(1): 5-16, Windhoek.

Vogel, J.C. & Rust, U. (1990): Ein in der Kleinen Eiszeit (Little Ice Age) begrabener Wald in der nördlichen Namib. - Berliner geogr. Studien 30: 15-34.

Vogel, J.C. & Visser, E. (1981): Pretoria radiocarbon dates II. - Radiocarbon 23: 43-80.

Waibel, L. (1922): Winterregen in Deutsch-Südwest-Afrika. - Abh. aus dem Gebiet der Auslandskunde, Vol. 9, Reihe C (Naturwiss.), Bd. 4, Hamburgische Universität, 112pp + appendix.

Zawada, P.K. (1997): Palaeoflood hydrology: method and application in flood-prone southern Africa. - S. Afr. J. Science 93: 111-132.

Palaeoenvironmental Transitions Between 22 ka and 8 ka in Monsoonally Influenced Namibia

Bernhard Eitel[1], Wolf Dieter Blümel[2] and Klaus Hüser[3]

[1] Geographisches Institut, Ruprecht-Karls-Universität Heidelberg,
Im Neuenheimer Feld 348, 69120 Heidelberg, Germany
Email: Bernhard.Eitel@urz.uni-heidelberg.de

[2] Institut für Geographie, Universität Stuttgart,
Azenbergstr. 12, 70174 Stuttgart, Germany

[3] Fachbereich Geowissenschaften, Universität Bayreuth,
Universitätsstr. 30, 95447 Bayreuth, Germany

Abstract

The paper presents a preliminary reconstruction of the development of different palaeoenvironments between the Last Glacial Maximum (LGM; c. 22 - 18 ka) and the Holocene Altithermal (HA; c. 8 ka - 4 ka) in Namibia. The synopsis is based on 36 optical datations of dune sands and fine-grained, silty deposits (OSL and TL). Most of the data were published by different research groups during the last decade. The synoptic view of all available optical age determinations is necessary because palaeoclimatic interpretations for southwestern Africa are not possible using results based only on local studies and on partly unreliable datations (e. g. ^{14}C ages of calcretes).

The compilation of all available datations and a synoptical interpretation such as the one presented here, show that gradual transitions and not abrupt changes from arid to more humid conditions occurred. These transitions did not affect all regions of Namibia at the same time and intensity. Differentiations in time and space are necessary for arriving a consistent model of the palaeoenvironmental transitions between LGM and HA.

The data compiled confirm previous results that during the LGM aridity affected all parts of Namibia, possibly with the exception the Kaprivi strip (northeastern Namibia) because dune fixation provides evidence that in the northern Kalahari basin desert changed to semiarid savanna conditions between c. 22 ka and 21 ka. At present this region receives 500 to more than 700 mm of mean annual rainfall. During the Mid-Holocene it was most probably slightly more. During the Late Pleistocene and Early Holocene the monsoonal influence increased, moister conditions shifted further south and west and at 14 ka reached the Windhoek highlands where savanna environments were established. In the dry transitional zone from the northern Namib Desert to the monsoonal Namibian highlands weak climatic oscillations are documented in sediments and soils. These oscillations, between c. 21 ka and c. 9 ka, agree with vegetation patterns which are suggested on the basis of marine core palynological studies off the Kunene River mouth for the Late Pleistocene and Early Holocene. In Namibia the most humid period since the LGM started approximately at 9-8 ka when the southwestern Kalahari, the arid core area of the intracontinental basin, changed to a savanna environment. At the same time monsoonal, more humid climates advanced against the eastern margin of the Skeleton Coast Desert (northern Namib Desert). After approximately 4 ka somewhat drier environmental conditions prevailed over the humid phase, and only minor hygric oscillations occurred until the present.

Zusammenfassung

Der Beitrag stellt einen Versuch dar, die natürlichen Umweltveränderungen zu rekonstruieren, die sich vom letzten Hochglazial (ca. 22 – 18 ka) bis zum holozänen Wärmeoptimum (ca. 8 – 4 ka) in Namibia vollzogen. Die synoptische Betrachtung basiert auf einer Zusammenstellung aller 36 bisher von verschiedenen Arbeistgruppen publizierten (und unpublizierten) optisch bestimmten Sedimentationsalter von Dünensanden und schluffigen Feinsedimenten (OSL, TL). Die Notwendigkeit einer synoptischen Betrachtung ausschließlich optisch datierter Sedimente erwächst aus der Erkenntnis, daß die paläoklimatische Interpretation für das südwestliche Afrika auf der Basis separater lokaler Studien und teilweise unsicherer Altersbestimmungen (z. B. ^{14}C-Alter von anorganischen Carbonaten) kaum möglich ist.

Die Zusammenstellung aller verfügbaren Sedimentalter (TL, OSL) in diesem Beitrag zeigt, daß sich der Übergang von hochglazial-ariden zu feuchteren Umweltbedingungen im mittleren Holozän allmählich und nicht abrupt vollzog, und daß dieser Prozeß nicht alle Regionen in Namibia zur gleichen Zeit und mit der selben Intensität erfaßte. Raum-zeitliche Differenzierungen sind notwendig, um ein widerspruchsfreies Bild der Paläoumweltentwicklung von der letzten Maximalvereisung der Erde bis zum mittelholozänen Wärmeoptimum entwerfen zu können.

Die zusammengestellten Daten bestätigen frühere Ergebnisse, nach denen nahezu das gesamte Namibia während des LGM arid war. Im nordöstlichen Namibia (Kavango/Kaprivi) wandelten sich wohl schon zwischen ca. 22 ka und 21 ka die wüstenartigen Bedingungen zu semiariden Savannen. Gegenwärtig erhält die Region jährlich ca. 500 bis über 700 mm Niederschlag. Im mittleren Holozän dürfte es zeitweise sogar noch etwas mehr gewesen sein. Im Spätpleistozän und frühen Holozän intensivierte sich der monsunale Einfluß und feuchtere Bedingungen reichten immer weiter nach Süden und Westen. Das Windhoeker Hochland wurde davon spätestens um 14 ka erfaßt und ist seit damals semiarid. Während des selben Zeitraums wurde die weiter anhaltend trockene Übergangszone von der nördlichen Namib zum monsunal geprägten Hochland Namibias von schwachen hygrischen Fluktuationen erfaßt, die in Sedimenten und Böden dokumentiert sind. Diese schwachen Fluktuationen zwischen ca. 21 ka und 9 ka passen gut zu palynologischen Auswertungen von Tiefseebohrkernen, die vor der Kunene-Mündung gezogen wurden. Die feuchteste Phase seit dem LGM in Namibia begann etwa vor 8000 bis 9000 Jahren, als sich in der südwestlichen Kalahari, die die trockene Kernzone des innerkontinentalen Beckens bildet, eine Savanne entwickelte. Zur gleichen Zeit erreichte der sommerfeuchte Monsun dauerhaft auch den heutigen Ostrand der nördlichen Namib (Skelettküstenwüste). Ab ca. 4 ka herrschen wieder etwas trockenere Umweltbedingungen, die seither nur noch von schwachen hygrischen Fluktuationen modifiziert wurden.

1 Introduction

A west-east transsect through southern Africa reminds us of the cross section of a tilted plate. The geomorphology of Namibia is characterized by a convex surface 80-120 km wide extending from the coast to the Great Escarpment from where the central highlands are inclined towards the Kalahari basin. East of the escarpment ephemeral rivers endoreically drain into this intracontinental basin (900 – 950 m a.s.l.) forming shallow omiramba (northwestern Kalahari) or incised mekgacha valleys (southwestern Kalahari). Neotectonics subdivide the Kalahari desert into basins and ridges (THOMAS & SHAW 1991, EITEL 1996). In northern Namibia the most prominent basin is the Ovambo basin with the Etosha Pan. In southeastern Namibia it is the Nama basin. The only perennial rivers which discharge into the Atlantic Ocean are the Orange River, which separates Namibia from South Africa, and the Kunene River at the western Angolan-Namibian frontier (Fig.1).

Southwestern Africa does not belong to a homogeneous climatic zone. Three climatic systems influence the area: episodic winterly rains from the South African Subtropics, the Benguela upwelling current in the southwestern Atlantic Ocean, and the Tropical monsoonal summer rainfalls.

At present, in Namibia only the region south of the Karas Mountains is sporadically affected by winter rainfall caused by westerly troughs north of the cape (TYSON 1986). The Namib Desert developed in response to the cool Benguela upwelling circulation and must be divided into three units: the outer part is a cold, foggy coastal desert (20-30 km), the middle part is a warm, foggy desert (30 km), and the eastern part is a warm inland desert with sporadic summer rains (BESLER

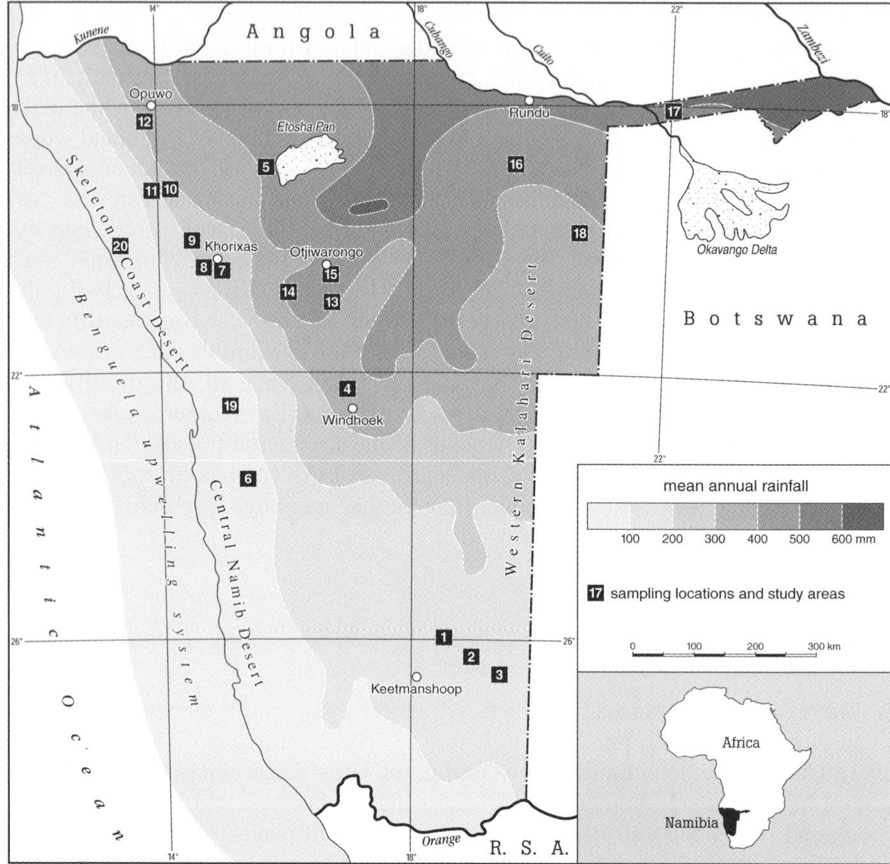

Fig. 1. Map of Namibia showing the mean annual rainfall at present and all study areas from where optical datations (OSL, TL) have been carried out until now. Note: The age of the sand deposition in the Rössing Cave indicates Benguela-induced Namib climates. In addition, it must be emphasized that the data are difficult to interpret due to methodological problems (HEINE 1992; locality 19). The datation of the Uniab River fan is only preliminary and methodological comments for a serious interpretation are not available (SCHEEPERS 1999; locality 20). Over and above that, the Uniab River fan fromation is strongly controlled by tectonic processes (HÜSER 1999, EITEL 1999) so that climatic deductions do not seem possible. The paper presented here concentrates on data representing different monsoonally induced climates east of the coastal desert. For sediment ages see Fig. 2.

1972). It is the only transitional zone to the semiarid Namibian highlands with monsoonal rains every summer. East of the Namib Desert the monsoonal influence decreases from north to south due to the seasonal shift of the Intertropical Convergence Zone (ITCZ). Therefore, the northern highlands are more humid than the southern regions, significantly decreasing from the highland savanna with more than 400 mm of mean annual precipitation to the hyperarid Skeleton Coast. In its southern parts the inner Namib Desert gradually merges, over a distance of 200 km, with the southwestern Kalahari Desert (150 mm mean annual precipitation; Fig. 1).

The paper focuses on monsoonally influenced Namibia which is – in somewhat simplified terms - the western part of the Kalahari basin. Sedimentologically, the western margin of the Kalahari is indicated by the outcrop of Tertiary Kalahari Group sediments (e.g. the Weissrand escarpment in southern Namibia) forming a prominent geomorphic lineament. For the present palaeoenvironmental purpose, the border of the monsoonal type `Kalahari climates´ lies in the transition zone and the eastern Namib Desert, including the Great Escarpment (250-450 mm mean annual precipitation). The long period of (semi-) aridity in southwestern Africa is related to the formation of the Kalahari Basin and the coastal Namib Desert. Except for the northern tropical savanna regions, southwestern Africa is dominated by desert, semi-desert and dry shrub or thornbush savanna.

It is generally agreed that the aridification of the western Kalahari basin began at least with the ice shield growth of Antarctica, the formation of the offshore Benguela upwelling circulation and the development of the Namib Desert during the Miocene (e.g. WARD ET AL. 1983; BESLER 1991). In southwestern Africa arid environments with endorheic drainage systems had their largest extent during the Upper Miocene and Pliocene. This is documented by the calcretization of the upper layers of the Kalahari Group deposits which buried nearly the whole of southwestern Africa as far north as the Kongo Basin (Mega Kalahari phase, THOMAS 1987; major calcrete generation, EITEL 1993; NETTERBERG 1969; VAN DER WATEREN & DUNAI 2001). Pedogenic palygorskite formation indicates clearly dry conditions in the Late Miocene and Lower Pliocene (EITEL 1994, 2000). Neither shelf sediments nor terrestrial deposits provide evidence of prominent humid periods in Namibia since the Pliocene. Soils (e.g. polygenetic calcretes), sediments and landforms clearly indicate only phases of weak hygric fluctuations. This has serious consequences for palaeoenvironmental research because palaeoclimates fluctuated only between desertic and dry semiarid conditions, and it is very difficult to identify and to date these palaeoenvironmental transitions: The soils are more or less calcareous. The carbonate prevents more intensive soil weathering and soil differentiation. Dryland soils contain very small amounts of humified organic matter, and desert soils are often completely free of soil organic carbon. Furthermore, fossils are very rare because termites eat wood and predators the bones of dead animals. This reduces the possibility to use soil organic matter for the absolute dating of Late Quaternary environmental changes. In contrast to the Saharan environments, it is more difficult to apply artefacts or paintings as palaeoenvironmental markers, because in southwestern Africa the neolithic period, handed down by the Khoisan people, lasted until the last few centuries.

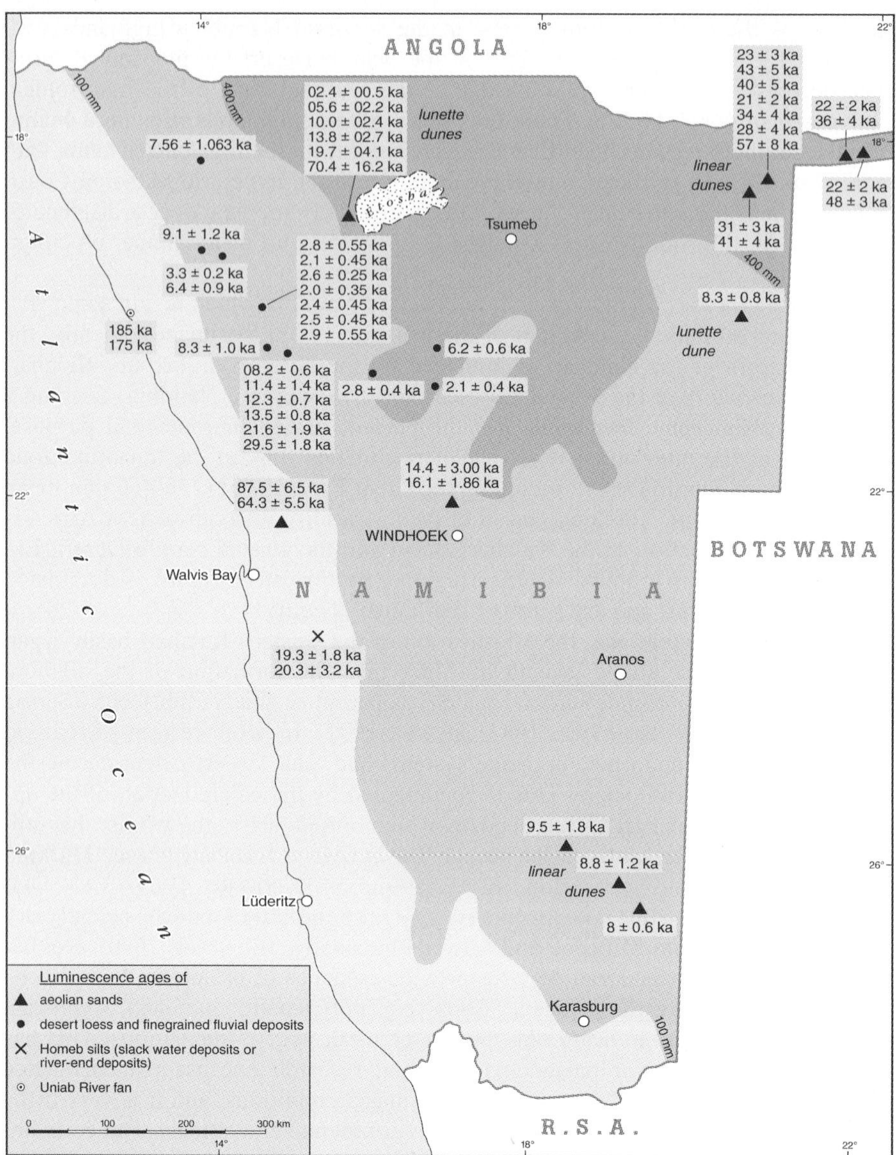

Fig. 2. Map with all the optical datations (TL, OSL only for Namibia; ages published and unpublished until now) presented by different authors (locations see Fig. 1, references see text).

Therefore, up to now Quaternary paleoclimatic reconstructions for southwestern Africa (cf. TYSON 1986; PARTRIDGE 1997, 1999) have remained highly speculative, due to a very small number of reliable absolute age determinations of sediments and soils. HEINE (1998) compiled a lot of data for the Namib Desert and referred to the difficulties of interpretation due to different methods of dating sediments and soils. Confusion derives especially from older attempts to determine sediment ages by radio isotopes of secondary carbonate. Dating Pleistocene calcretes (by radiocarbon isotope analysis, Th/U isotope ratios or ESR) is mostly not reliable due to multiple carbonate recrystallizations, leaching effects etc. (e.g. GEYH & EITEL 1998). In addition, dryland geomorphodynamics remove especially fine-grained sediments and soil particles, distribute them over a more or less extended area, and incorporate them into younger deposits (BLÜMEL 1982, EITEL 1993). The recycling passes older sedimentary or pedological features onto younger formations. This explains the difficulties of using soil and mineral charcteristics for revealing Pleistocene environmental change.

Because of these problems, for about ten years new attempts have been made using optical dating of fine-grained sediments for investigating Late Quaternary paleoenvironmental change east of the Namib Desert. The application of this method was accompanied by geomorphological and pedological studies of dune sands, soils, alluvial silts and similar loessic deposits in different regions of Namibia.

The LGM has been identified not only as a cool, but also as a dry period in deserts, and the Holocene warming period culminated in the HA, 8-4 ka. In most modern drylands this period is characterized by more humid conditions (SARNTHEIN 1978). For most parts of southern Africa it is commonly accepted that the LGM was dry and the Mid-Holocene moist (e.g. PARTRIDGE 1999). But little is known about the transition from one to the other of those very different climatic states east of the Namib Desert. The present paper compiles published and unpublished geomorphological data from optically dated sediments, trying to create a reliable foundation for a chronology of the environmental transition from the LGM to the HA in the monsoonally influenced parts of Namibia. The synopsis focuses on the transition from deserts (c. 22 ka ago) to semiarid savannas (c. 8 ka ago), and on monsoonal effects on the eastern margin of the coastal desert. Therefore, we will not discuss TL and/or OSL data from beyond this time span, and sediments indicating only Namib palaeoenvironments controlled by locally reduced or increased Benguela upwelling (see Fig. 1, location 19: Rössing cave, HEINE 1998) or by local tectonics (see Fig. 1 location 20: Uniab River fan, SCHEEPERS & RUST 1999) will be ignored.

2 Data from dune sands

During the last decade luminescence ages of dune sands have been obtained from the southwestern Kalahari basin, the Windhoek highlands and the northern regions of Namibia.

2.1 Dune shifting in the southwestern Kalahari (Namibia)

At present the southwestern Kalahari receives a mean annual precipitation of less than 200 mm. It is the driest part of the innercontinental basin. During the Tertiary the Proto-Fish, the Proto-Auob and the Proto-Nossob River filled the Nama basin with more than 200 m of fluvial deposits belonging to the Kalahari Group. The top layers are calcretizised and belong to the Tertiary calcrete generation (cf. WATTS 1980, EITEL 1993). On the margins of the basin the Kalahari Group deposits are less than 50 m thick and numerous pans formed on it during the Quaternary. Linear dunes 5 to 15 m high cover some older pans or cross them, confirming drier periods of reduced pan formation and increased dune shifting. Sedimentological and mineralogical studies all show that the dune sands are derived from the Tertiary sediments of the Nama basin and not from the Karoo sand-/siltstones below. The reddish colour of the sands is not a sign of Quaternary humid conditions but it has been inherited from the Upper Tertiary layers (EITEL & BLÜMEL 1997).

For the Namibian part of the southwestern Kalahari the only luminescence data available so far were published by BLÜMEL et al. (1998). The data show that linear dune shifting ended between 9.5 ka and 8 ka in the Keetmanshoop-Tses-Aroab triangle (Fig. 1, sampling locations 1-3). The ages accord with data from southern Botswana and South Africa where linear dunes were found to have been active from 16 to 9 ka (STOKES et al. 1997).

2.2 Aridity in the Windhoek highlands: The age of the Teufelsbach sandfield

In the Windhoek highlands, approximately 20 km south of Okahandja, the main road B1 crosses a dune field of several km² extension (location 4, Fig. 1). The Teufelsbach railway station is the nearest well-known site. The sandfield consists of different dune types caused by the broad valley floor and by varying wind regimes in the adjacent mountainous region. The maximum height of the dunes is approximately 7 m. At present the dunes are fixed by vegetation, and only some crests are active because due to overgrazing by cattle. In 1996 road works crossed some of the dunes, and thereby giving direct access to the consolidated dune base. The presence of undisturbed stratification excludes any postsedimentary reworking (e.g. by termites).

Sand samples ca. 1 m above the dune base from two neighbouring dune complexes were dated optically. The age of dune building has been determined to be 14 ± 3 ka (lab-no. HDS 649; Table 1). This accords with the age of the second dune complex which was determined by OSL (13.79 ± 3.07 ka) as well as by TL (13.86 ± 3.31 ka; lab. Geogr. Inst. Univ. Köln).

2.3 The Okondeka lunette dune formation

The first TL data from dune sands of northern Namibia came from the western rim of the Etosha Pan (BUCH et al. 1992). Eight samples were taken near Okondeka (Fig. 1, location 5). The age determination of the sand deposition indicates different periods of increased aeolian activity over the last 140 ka, namely at 70.4 ± 16.2 ka, at 19.7 ± 4.1 ka, at 14.8 ± 2.8 ka, at 13.8 ± 2.7 ka, 10.0 ± 2.4 ka during the Latest Pleistocene, and during the Mid-Holocene at 5.6 ± 2.2 ka and 2.4 ± 0.5 ka (BUCH et al. 1992).

The data are of special interest because they do not only indicate periods of lunette dune formation. They also indicate increased deflation from more than 5,600 km² pan surface within the boundaries of the Etosha National Park (LINDEQUE & ARCHIBALD 1991). Large quantities of windblown dust originating from the western Kalahari Desert have been blown out and accumulated in basins and valleys of the eastern Namib Desert. Therefore, special attention was paid to silty deposits on the eastern margin of the coastal desert (see section 3).

2.4 Dune activity in the northern and northwestern Kalahari (Namibia)

In the northern Kalahari basin linear dune systems show an east-west orientation which is believed to provide evidence of palaeowind systems (eg. LANCASTER 1981). At present northeastern Namibia (Kavango and Caprivi) receives more than 450 mm of mean annual rainfall and the dune systems are fixed by savanna woodlands. Age determinations (OSL) of dune sands from four sites (Fig. 1, location 16) south of Rundu indicate more or less continuous dune construction between c. 43 and 21 ka, while the ages from two sites in Caprivi (Fig. 1, location 17) provide evidence for dune activity at c. 48 ± 3 ka, 36 ± 4 ka and 22 ± 2 ka

Table 1. Sediment parameter and OSL data from a dune in the central Teufelsbach sandfield (Windhoek highlands). The data concur with OSL and TL data from dune sands in the vicinity (s. text).

Grain sizes [mm]	<0.002	0.002 - 0.006	0.006 - 0.02	0.02 - 0.06	0.06 - 0.2	0.2 - 0.63	0.63 - 2	>2
HDS 649 [%]	3.19	0	0.97	5.06	77.41	13.31	0.07	0
$CaCO_3$		0.2%			Munsell colour		10YR4/6	

Lab.No.	D	U [μg/g]	Th [μg/g]	K [%]	DE_β [Gy]	Total Dose rate [Gy ka^{-1}]	OSL Age [ka], 1σ error
HDS 649	1.04 ±0.04	3.1 ±0.01	9.63 ±0.20	2.01 ±0.06	49 ±11	3.5 ±0.06	14 ±3

(THOMAS et al. 2000). It should be noted that, in contrast to the linear dune formation in the southwestern Kalahari and in the Windhoek Highlands, dune fixation coincides with the beginning of the LGM.

One further date has been published by HEINE (1995) from a lunette dune west of Nyae Nyae Pan, 15 km south of Tsumkwe in Bushmanland/northeastern Namibia (Fig. 1, location 18). The age of the lunette dune has been determined by TL to be 8.3 ± 0.8 ka. On the one hand, the date differs considerably from the results by THOMAS et al. (2000) from dunes in northeastern Namibia, on the other hand there seems to be a correspondence with the period of dune fixation approximately 750 km further south in the southwestern Kalahari (EITEL & BLÜMEL 1997, BLÜMEL et al. 1998). It should be noted, that HEINE (1995) studied a lunette dune system, whereas the other authors dated linear dunes. This helps explain the discrepancies between the results and makes clear that there are different geomorphogenetic processes responsible for the sedimentation. No direct palaeoclimatic link exists between linear dune fixation and lunette dune formation.

3 Data from silty deposits

Fine-grained silty deposits have been found and investigated in the Namib Desert and in the eastern transitional zone to the Namibian highlands. A compilation of locations of river silt terraces in northwestern Namibia was presented by RUST (1987, 1989b). The sediments fill basins and deep valleys more than 10 m deep. Sediment analysis shows that the silts are aeolian materials (mainly Kalahari dust) and some fine-grained local weathering products (EITEL et al. 1999b).

3.1 The Homeb silts

At Homeb (middle Kuiseb River valley / central Namib Desert; Fig. 1, location 6) remnants of laminated silt deposits occur which filled the Kuiseb valley more than 20 m deep. The Homeb silts formed between 20.3 ± 3.2 ka (base) and 19.3 ± 1.8 ka (top) (TL data; EITEL & ZÖLLER 1996). These TL ages concur with ^{14}C datations (23 - 19 ka BP; VOGEL 1982).

3.2 Silts in the Aba-Huab and Huab River catchment

South of Khorixas (northern Damaraland) fine-grained silty deposits fill more or less extended basins in the catchment of the upper Aba-Huab River, the southern branch of the Huab River. At present the region receives about 200 mm ofmean annual precipitation. The large west-east orientated Aba-Huab basin probably formed during the Palaeozoic (MARTIN 1969). The silty fillings are approximately 5 m thick, formed by the fluvial deposition of mainly aeolian materials and local weathering detritus during the Late Pleistocene and Early Holocene. At present, on Dieprivier Farm and Uitskot Farm (Fig. 1, location 7) the uppermost Aba-Huab tributary dissects the floor of the basin, thereby causing gullies in which the structure of the deposit becomes exposed.

Thermoluminescence data show that the sedimentation started at 29.5 ±1.8 ka. New OSL data confirm this date by a minimum age of 25 ± 4 ka and a maximum age of 29 ±4 ka (lab.-no. HDS 647, Table 2). Further sedimentation took place at 21.6 ±1.9 ka, and from 13.5 ± 0.8 ka to 12.3 ± 0.7 ka. The final deposition was at 8.2 ± 0.6 ka, followed by fluvial dissection and erosion. Sedimentation was interrupted by four periods of geomorphic stability with weak soil formation (calcretization, pedogenic Fe dynamics) (EITEL & ZÖLLER 1996).

Dieprivier is the only site where the age of the sediment-soil sequence has been studied by various methods (TL, ^{14}C and Th/U). The results of the radio isotope analysis range widely and/or are not consistent with the field observation. This confirms former results that the dating of Pleistocene carbonates (soil formation) by radio isotopes is mostly not applicable, due to repeated carbonate recrystallization and leaching effects (BLÜMEL 1981). At best it seems possible to date very young calcretes (GEYH & EITEL 1998).

Some km further west on Inhoek Farm (Fig. 1, location 8) another small branch of the upper Aba-Huab River dissects the basin fill to a depth of more than 3 m. The uppermost layer of the exposed fine-grained deposits has an age (OSL) of 8.3 ± 1.0 ka (EITEL et al. 2001). This corresponds with the age of the final deposition at Dieprivier Farm and Uitskot Farm.

In the upper Huab River valley, approximately 0.5 km west of the Huab Lodge (870 m a.s.l.; ancient Garubib Farm), desert loess has been found (Fig. 1, location 9). At present the region receives ± 250 mm/a rainfall, supporting an open mopane savanna. Two prominent terraces are developed in the valley, one at 2.5 m, the other one 4 m above the present river bed. The 4 m terrace is gently inclined towards the Huab River. It ascends over a distance of approximately 400 m to the northern slope of the valley. Colluvial weathering detritus, short-transport angular quartz pebbles up to coarse breccia size of 50 cm, has been laterally incorporated into the alluvial deposit. Aeolian sediments have been deposited onto this alluvial and colluvial sediment complex. They are 2 m thick, forming a rolling surface caused by different accumulation rates and soil erosion. The desert loess is partly sandy and shows no sedimentary or textural signs of colluvial or alluvial superimposition. Runoff from the backward slope cannot reach the fine-grained deposit because deep erosion channels have dissected the terrace and lead directly to the present talweg.

The OSL data show that the aeolian sediment was deposited in the Holocene, at about 2-3 ka (Table 3). This was most likely caused by increased deflation in the northwestern Kalahari basin. The exact age determination is difficult because of weak soil formation and the high total dose rate.

Fig. 3. Basins in upper Aba-Huab catchment south of Khorixas. They are filled with fine-grained deposits (predominantly silt) from aeolian Kalahari dust and local rock weathering. Note: The slopes are covered by coarse clasts. Weak runoff has washed the particles from the slopes into the basin, accumulating c. 5 m of silty deposits. Above: View of a sequence of small basins filled with silts. Below: One of the basins in the vicinity of Dieprivier Farm illustrating actual gully erosion.

Table 2. OSL data from the basal layer in the Aba-Huab basin (Dieprivier Farm / Khorixas District). Minimum and maximum age due to radioactive imbalance. The silt deposition started at the beginning of O-stage 2. This confirms former TL data presented by EITEL & ZÖLLER (1996).

Grain sizes [mm]	<0.002	0.002 - 0.006	0.006 - 0.02	0.02 - 0.06	0.06 - 0.2	0.2 - 0.63	0.63 - 2	>2
HDS 647 [%]	24.17	7.72	15.35	24.77	24.27	2.91	0.80	0
$CaCO_3$		21.4%			Munsell colour		10YR4/4	

Lab.No.	D	U [µg/g]	Th [µg/g]	K [%]	DE_β [Gy]	Total Dose rate [Gy ka^{-1}]	OSL Age [ka], 1σ error
HDS 647	1.04 ±0.04	2.58 ±0.06 3.64 ±0.35	11.89 ±0.18	2.20 ±0.06	119.6 ±16.6	4.14 ±0.26 4.84 ±0.38	(min) 25 ±4 (max) 29 ±4

3.3 Silts in the upper Hoanib River catchment

Coming down from the northwestern Kalahari rim (Etosha), the Ombonde-Hoanib River crosses north-south orientated basins and mountainous ridges in epigenetic valleys and gorges in the southern Kaokoveld (BRUNOTTE & SPÖNEMANN 1997). In the headwater basins fine-grained, mainly silty deposits are widespread, which are more or less clayey and, depending of the amount of calciumcarbonate, light brownish to white. A typical example are the dusty deposits in the more than 10 km wide Ombonde-Aap basin (Ombonde Vlakte, Fig. 1, location 10) east of the Khowarib gorge, upper Hoanib River catchment (Fig. 1, location 11). First preliminary datations of weakly consolidated surficial layers of the more than 8 m thick Ombonde Vlakte silts provide evidence of different sedimentation phases, at 3.3 ± 0.2 ka (lab.-no. HDS 646) and at 6.4 ± 0.9 ka (lab.-no. HDS 648), indicating repeated periods of dust flux from the Etosha region during the Holocene (OSL ages, Table 4). Further systematical sampling of these fills was carried out in October 2001. At the present stage of investigations, a definitive interpretation of the deposits is not yet possible.

Table 3. OSL data and sedimentological characteristics of the aeolian sediment on the 4 m terrace at Garubib Farm (Fig. 1, location 9). The desert loess was deposited after the HA, most likely between 3-2 ka (EITEL ET AL., in prep.).

Grain sizes [mm]	<0.002	0.002 - 0.006	0.006 - 0.02	0.02 - 0.06	0.06 - 0.2	0.2 - 0.63	0.63 - 2	>2
D 18 [%]	2.9	3.2	12.4	30.2	39.1	10.4	1.8	0
$CaCO_3$		3.8%			Munsell colour		10YR4/6	
D 19 [%]	11	9.1	22.1	34.1	22.2	1.3	0.2	0
$CaCO_3$		2.9%			Munsell colour		10YR4/4	
D 20 [%]	5.4	4.6	17.5	49.4	22.2	0.8	0.1	0
$CaCO_3$		1.7%			Munsell colour		10YR4/6	
D 21 [%]	5.7	5.7	16	21.2	35.1	13.8	2.5	0.4
$CaCO_3$		1.5%			Munsell colour		10YR4/6	
D 22 [%]	8.6	4.1	15.9	40.8	26.8	3.2	0.6	0.4
$CaCO_3$		1.2%			Munsell colour		10YR4/4	
D 23 [%]	14.2	12.5	25.2	29.7	16	2	0.4	0
$CaCO_3$		1.7%			Munsell colour		10YR4/4	
D 25 [%]	14.3	11.1	23.9	29.4	17.6	2.8	0.9	0.1
$CaCO_3$		1.4%			Munsell colour		10YR4/4	

Lab.No.	D	U [µg/g]	Th [µg/g]	K [%]	DE_β [Gy]	Total Dose rate [Gy ka^{-1}]	OSL Age [ka], 1σ error
D 18	1.1 ±0.1	2.65 ±0.05	14.2 ±0.45	3.3 ±0.09	16.1 ±1.2	5.82 ±1.06	2.8 ±0.55
D 19	1.01 ±0.01	3.43 ±0.06	15.7 ±0.52	3.25 ±0.13	15.4 ±2.9	7.2 ±0.34	2.1 ±0.4
D 20	1.01 ±0.01	3.12 ±0.06	13.8 ±0.5	2.61 ±0.1	16.2 ±1.3	6.37 ±0.29	2.6 ±0.2
D 21	1.01 ±0.01	3.17 ±0.06	12.9 ±0.42	3.19 ±0.13	13.4 ±1.8	6.91 ±0.33	2.0 ±0.3
D 22	1.01 ±0.01	2.43 ±0.06	12.8 ±0.27	2.80 ±0.11	14.5 ±2.2	6.14 ±0.28	2.4 ±0.4
D 23	1.01 ±0.01	2.25 ±0.15	20.6 ±0.7	2.3 ±0.09	17.9 ±2.6	6.08 ±0.30	2.5 ±0.4
D 25	1.01 ±0.01	1.9 ±0.13	15.8 ±0.3	2.2 ±0.07	18.6 ±3.0	6.41 ±0.31	2.9 ±0.5

Table 4. OSL data from the Ombonde Vlakte silts / Upper Hoanib River (NW-Namibia).

Grain sizes [mm]	<0.002	0.002 - 0.006	0.006 - 0.02	0.02 - 0.06	0.06 - 0.2	0.2 - 0.63	0.63 - 2	>2
HDS-646 [%]	14.52	20.03	30.31	25.20	7.22	2.12	0.6	0
$CaCO_3$		0.2%			Munsell colour		10YR3/3	
HDS-648 [%]	18.45	16.49	23.0	20.69	14.25	4.61	2.51	0
$CaCO_3$		1.2%			Munsell colour		10YR4/3	

Lab.No.	D	U [µg/g]	Th [µg/g]	K [%]	DE_β [Gy]	Total Dose rate [Gy ka^{-1}]	OSL Age [ka], 1σ error
HDS-646	1.04 ±0.04	4.09 ±0.06	17.23 ±0.18	3.10 ±0.05	23.1 ±0.7	6.98 ±0.04 6.52 ±0.38	(min) 3.3 ±0.2 (max) 3.5 ±0.2
HDS-648	1.04 ±0.04	3.1 ±0.9	13.46 ±0.34	2..45 ±0.11	34.1 ±4.1	5.31 ±0.28	6.4 ±0.9

In the Khowarib gorge (Fig. 1, location 11) two distinct silt complexes occur, the white Khowarib-1 terrace (c. 15 m high) of unknown age, and the brownish Khowarib-2 terrace (c. 5 to 6 m height). Three younger, less prominent terraces up to 3 m high formed after the Khowarib-2 complex. They are sandy with intercalated silty layers, having been formed by episodic flash floods. In the middle of the gorge the brownish terrace crosses onto the Khowarib-1 terrace, serving as a good example of terrace intersection. The top layer of the Khowarib-2 silt terrace was deposited at 9.1 ± 1.2 ka (EITEL et al. 2001).

3.4 Silts in the upper Hoarusib River catchment

Southwest of Opuwo in the Hoarusib River catchment (Fig. 1, location 12) fine-grained sediments occur which are similar to the deposits in the headwater basins of the Aba-Huab and Hoanib rivers mentioned above. They are covered by a mid-Holocene loesslike aeolian (?) deposit. A first datation of the basin fill underneath them shows that sedimentation lasted until 7.56 ± 1.063 ka (BRUNOTTE & SANDER (2000). A single age determination is not very reliable, but the end of the deposition at about 8 ka concurs with the time when the accumulation of similar fills ended at the Dieprivier/Uitskot Farms, the Inhoek Farm, similarly deposition of the Khowarib-2 terrace ended and erosion began.

3.5 Holocene deposits in the Otjiwarongo thornbush savanna

Pedological and environmental studies in the Otjiwarongo region show that late Holocene dark soils, e.g. Calcisol-Vertisol-Kastanozem associations formed in Mid-Holocene fine-grained deposits (EITEL & EBERLE 2001). At 1,400 m to 1,600 m a.s.l. the sediments fill shallow valleys and basins on both sides of the flexural bulge separating the uppermost Ugab catchment from the Omatako Omurambo system, discharging endorheically into the Kalahari basin. The deposits occur as fine-grained, mainly fluvial layers 0.5 to 2 m thick covering calcretes of unknown age or directly overlying the basement rocks.

East of the Omatako Mountains, on Farm Wewelsburg (Fig. 1, location 13) the OSL age of the layer is 2.1 ± 0.4 ka (HDS 795). West of Kalkfeld on Osongombo-Ost Farm (Fig. 1, location 13) it is 2.8 ± 0.4 ka (HDS 796). In the Otjiwarongo townlands (Fig. 1, location 14) a similar layer covers the calcrete on top of a hill. The OSL age is 6.2 ± 0.6 ka (HDS 794) (EITEL et al. 2002).

More data are necessary to assess the paleoclimatic relevance of these fine-grained Holocene cover beds. Their formation however, supports minor geomorphodynamic changes following the Holocene Altithermal.

4 Discussion and synoptic interpretation

It has been argued that dune formation depends on channel flow and fluvial sediment availability and that dunes do not necessary provide evidence of pronounced aridity at the time of dune formation, but of fluvial activities and sediment supply (WILLIAMS 1986). However, unequivocal geomorphological and sedimentological evidence exist that the Kalahari dune systems (up to 100 km in length) and hugh quantities of dust blown to western Namibia mainly derive from the aeolian reworking of weathered and mostly calcretizised Kalahari Group deposits (Tertiary) (EITEL 1994, EITEL & BLÜMEL 1997). Dune shifting and the formation of sand fields are assumed to be clear indicators of arid conditions because both require extended gaps in the plant cover.

Linear dune systems have been used to deduce prevailing winds assumed to coincide with the dune orientation. In the Kalahari LANCASTER (1981, 1989) and THOMAS & SHAW (1991) discussed the possibility that the oriented linear dunes provide evidence of stable circulation patterns over southern Africa. It is difficult to verify the relationship between linear dunes and special atmospheric constellations though, because nobody knows when and under which conditions the orientation took place. Pans as well as linear dunes are due to complex geomorphic processes (e.g. HEINE 1990; LIVINGSTONE & THOMAS 1993) and have a polygenetic and polyclimatic heritage (draa-type dunes; BESLER 1992).

As for dune mobilization, LANCASTER (1988) suggested that, given a 5 K mean temperature decrease and a wind velocity increase of 117% of the present mean (after NEWELL et al. 1981), a fall in precipitation of 30 - 40 % would be sufficient. This matches with, for example, the presumed decrease to 100 mm or even less of mean annual rainfall (100 - 200 mm/y today) over the southwestern Kalahari Desert during the LGM (PARTRIDGE 1997).

4.1 Late Pleistocene and Holocene environmental transitions in Namibia: Evidence from dune systems

During oxygene isotope stages 4 to 2 the aridity in northern Namibia was repeatedly weakened or even interrupted by more humid phases. This is indicated not only by dune systems in Caprivi (THOMAS et al. 2000), but also by lake level oscillations of Etosha (RUST 1984, BUCH 1996), `Lake Kaprivi´ (SHAW & THOMAS, 1988), `Lake Paleo-Makgadikgadi´and Lake `Ngami-Okavango´ (SHAW & COOKE 1986; THOMAS & SHAW 1991). More proxy data from the northern and central Kalahari Desert were compiled and discussed in THOMAS & SHAW (1991), SHAW & THOMAS (1996) and STOKES et al. (1997).

It is generally accepted that the LGM produced the driest environments in southwestern Africa within the last 30 ka (e.g. TYSON 1986, PARTRIDGE 1999). Different times of dune fixation in southwestern Africa provide evidence of the idea that arid climates were followed by moister conditions which, controlled by the intensity of the monsoonal circulation, shifted from north-northeast to south-southwest up to the HA. The southwestern Kalahari is the `arid core´ of the inner-continental basin, and it is believed to be an environment where aridity started earlier and lasted longer, while the more prominent climatic changes in the middle and northern Kalahari basin favoured longer records of multiple phases of dune building (THOMAS et al. 2000).

1. In northern Namibia dune building in Kavango and Caprivi until c. 20 ka (THOMAS et al. 2000) and Etosha lunette dune formation c. 19 ka (BUCH et al. 1992) provide evidence of arid environments with increased aeolian activity. After 20 to 19 ka the desert conditions were transformed into semi-arid savanna-type environments.
2. At 14 ka late Pleistocene aridity ended in central Namibia. In the Windhoek highlands the age of the final fixation of the Teufelsbach sandfield is a clear signal of increasing monsoonal rainfalls in southwestern Africa up to the the Late Pleistocene. Prior to 14 ka the desert conditions in the Windhoek highlands suggest that the formation of the Homeb silts (during LGM) most likely occurred during a dry period (see section 4.2).
3. For the time after 14 ka the more humid conditions in the Windhoek highlands explain the flooding of the Auob and Nossob dry valleys (mekgacha) in the southwestern Kalahari (c. 11 ka; HEINE 1981) where aridity still lingered in the Late Pleistocene. It seems that a dry transitional belt persisted between the Namib Desert and the ancient Kalahari Desert, still reaching approximately 22°S during the Latest Pleistocene (Fig. 4).
4. The increasing monsoonal influence reached the southwestern Kalahari basin in the Early Holocene. Dune ages presented by STOKES et al. (1997) from sandfields in Botswana show final shifting at about 9 ka. In south-eastern Namibia dune shifting ended approximately 9 to 8 ka (EITEL &

BLÜMEL 1997). The coincidence of the ages of the linear dune activity over a distance of about 200 km suggests that dune fixation was not the result of local effects, but due to a subcontinental environmental change. The difference of 1 ka between beginning dune stabilization in the southwestern Kalahari basin can be explained either by delayed regrowth of the plant cover due to somewhat drier conditions in the arid core of the southwestern Kalahari or by the 1σ error in age calculation.

Since the HA semiarid conditions have persisted in the southwestern Kalahari because linear dunes have remained stable. Subsequent hygric fluctuations were weak, and until now there has been no evidence that Holocene climates went beyond the semiarid range.

4.2 Monsoonal effects on the eastern margin of the Namib Desert: Evidence from fine-grained silty deposits

It is more difficult to reconstruct the monsoonal effects of increasing moister conditions on environments at the eastern margin of the Namib Desert. Fine-grained deposits are widespread in northern Damaraland and Kaokoveld (see section 3). They are more or less homogeneous, mostly calcareous and in general they display specific features (e.g. lamination, thin stratification). It should be noted that there is some disagreement as to their interpretation, due to different types of silty deposits. Confusion exists because silt deposits occur as large basin and valley fills up to several metres thick or as small river terraces. Under discussion are genetic models such favouring high-energy flashflood (SMITH et al. 1993) or slackwater deposits (e.g. HEINE 1998), low-energy river-end deposits (eg. EITEL et al. 1999b; RUST 1999), or silts deposited behind dune barriers (e.g. BLÜMEL et al. 2000).

The discussion began with the interpretation of the Homeb silts (Fig. 1, location 6) in the 1970s. RUST & WIENEKE (1978) interpreted the sediments as dune dam (lake-) deposits due to increased dune shifting of the Namib sand sea in the vicinity. This would reflect more arid conditions in the river catchment (Khomas Highlands) because during sedimentation the ephemeral Kuiseb river had not been able to break through the dune dam. GOUDIE (1972) and MARKER & MÜLLER (1978), who presented the most detailed sedimentological study, excluded dams and interpreted the silts as low-energy river-end deposits and indicative of arid conditions as well, whereas OLLIER (1977) and HEINE (1998) favoured an interpretation as floodplain sediments or high-energy slackwater deposit of an aggrading Kuiseb river, indicating increased rainfall in its upper catchment.

The different origin of the silt deposits may explain their different ages, ranging from just a few 100 a to more than 40 ka (radiocarbon data and optical age determinations; e.g. RUST & VOGEL 1988; RUST 1989a; RUST & VOGEL 1990; RUST 1999).

Extended silty basin fills in the headwater catchments (c. 600 m – 1,200 m a.s.l.) are best interpreted palaeoclimatically for following reasons:

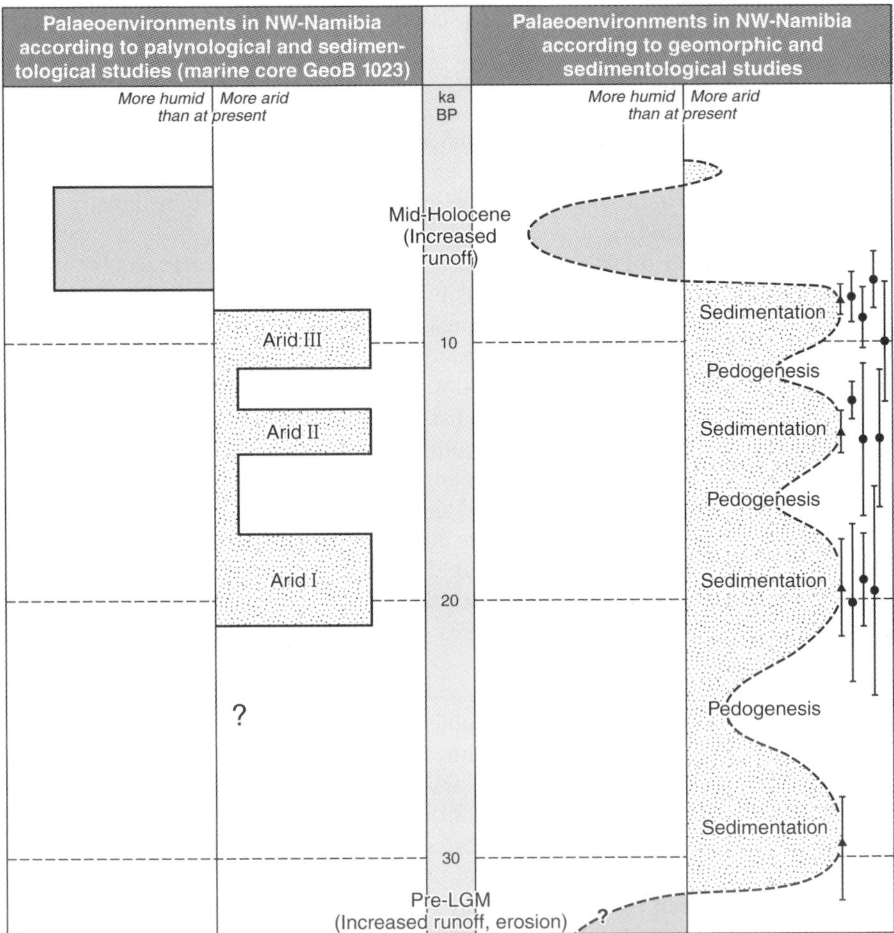

Fig. 4. Correlation of data from optically dated sediments (Late Pleistocene to Early Holocene) including 1σ error bars from the upper Aba-Huab River (Dieprivier Farm, bars with triangles) combined with the results of marine core palynological and sedimentological studies (left side). Further optical datations from sediments in northwestern Namibia (bars with dots; see locations 5, 6, 8, 9, 11, 12 in Fig. 1) confirm the two weak hygric oscillations in northwestern Namibia during an arid period which lasted here since the LGM until the Early Holocene. References and remarks in the text.

1. Dune damming is restricted to the coastal desert. The silts are not lake deposits because they follow the general valley and basin gradient.
2. Slackwater deposits are favoured further downstream, because headwater flash flood heights are smaller due to the high number of shallow and branching tributaries. For generating such deposits it is necessary for the height of the floods to increase with the concentration of the runoff some

km downstream and for related sediments to benefit fine-grained deposits removed from upstream.

3. The remobilization of silts explains more local river-end deposits of different age in various parts of the river courses down to the Skeleton coast, due to climatic events on a subregional scale (RUST 1999). In such cases the silt deposits are an expression of the local geomorphodynamics in drylands and not a sign of (sub-) continental climatic change in the monsoonally influenced hinterland.

The extended and thick fine-grained deposits in the headwater catchments of the Aba-Huab, Huab, Hoanib or Hoarusib River east of the Namib Desert (c. 150 – 250 mm of mean annual rainfall at present) were deposited over a long period due to changing intensities of the monsoonal regime. For this case a model has been developed comprising silt deposition caused by aridification with increased aeolian activity in the western Kalahari basin, dust impact on the eastern margin of the coastal desert, and reduced local runoff. The rainfalls were sufficient to wash the dust, together with fine-grained weathering products, from slopes and to accumulate them in headwater valleys and basins, but there was not enough runoff to carry the materials over longer distances out of the region to the central coastal desert or into the Atlantic Ocean (EITEL et al. 1999b). The typical feature of all fills is a knick between the surface of the silt deposits and the older slopes often covered with coarse boulders. This knick is confirmation of short-distance alluvial transport.

In the easternmost Skeleton Desert such conditions started with weak hygric fluctuations after approximately 29 ka and ended approximately 8 ka, as indicated by the final deposition of the fine-grained alluvium at Dieprivier Farm, Inhoek Farm and from the upper Hoarusib River and the Hoanib River canyon upstream of Khowarib. The assumed dust flux from the western Kalahari Desert to the Damaraland and Kaokoveld (northwestern Namibia) is confirmed by marine sediment records from the shelf off Namibia indicating increased aeolian impact on the Atlantic Ocean (SUMMERHAYS et al. 1995). Increased dust flux out of the western Kalahari basin does not contrast with the establishment of semiarid environments in the northwestern Kalahari basin after approximately 19 ka. Pan activity needs at least episodic rainfalls and is part of semiarid as well as of desert geomorphodynamics (GOUDIE & WELLS 1995, EITEL & BLÜMEL 1997). Even at present, dust storms can be frequent at the beginning of the rainy season.

The period between 20 ka and 8 ka in northern Namibia was characterized by weak climatic oscillations which are documented by soil-sediment complexes at Okondeka (BUCH et al. 1992) and in the upper Aba-Huab River basin at Dieprivier-Uitskot (EITEL & ZÖLLER 1996). The Okondeka-II soil complex developed at least during three phases until the end of the Pleistocene/earliest Holocene (BUCH et al. 1992). At Dieprivier Farm pedogenesis occurred approximately between 21.6 ka and 13.5 ka and from 12.3 ka to 8.2 ka (EITEL & ZÖLLER 1996). The buried soils are more or less indurated calcisols, partly with hydromorphic characteristics believed to provide evidence of geomorphic stability. In contrast to the Okondeka soil complexes formed in dune sands, the soils at Dieprivier developed in silts. These fine-grained sediments show no signs of erosion, such as buried stream channels, for the periods of pedogenesis. Therefore, runoff did not increase during these phases. Perhaps weak advective rainfall reached the desert margin, because there is no evidence of intensified convective monsoonal summer rains.

The idea of phases of weakened aridity within a dry period after the LGM matches very well the results of palynological studies from marine core GeoB 1023 off the Kunene River mouth. According to SHI et al. (2000) hygric fluctuations are documented in pollen diagrams indicating three more arid phases weakened by two more humid Late Pleistocene periods in the southwestern Angolan and northwestern Namibian highlands (Fig. 4).

The end of the dry phase at approximately 9 to 8 ka concurs with the fixation of the linear dunes in the southwestern Kalahari Desert, the results of the marine core GeoB 1023 palynological studies (SHI et al. 2000) and with increased discharge of the Kunene River into the Atlantic Ocean as recorded by the clay mineral record in the marine core GeoB 1023-4 (GINGELE 1997; Fig. 4). In the transition zone from the Namib Desert to the Namibian highlands the moister conditions favoured runoff and prominent silty river terraces were formed. The shift to more humid semiarid environments included nearly the whole Kalahari basin and is documented at several sites in southern Africa as well (PARTRIDGE 1997). In western Namibia the intensified monsoonal influence was stopped by increased Benguela upwelling which produced hyperarid conditions in the coastal desert (SUMMERHAYS et al. 1995). During the HA this opposite climatic regime led to a prominent ecological gradient from savanna-type environments to the hyperarid coastal desert (Fig. 5).

During the Mid-Holocene more humid conditions did not stop the aeolian dynamics and the deflational activity from the Etosha Pan oscillated, as indicated by lunette dune activity at Okondeka and by silt deposition in the Ombonde-Hoanib catchment. The work on this topic is going on by dating desert loess and palaeosoil sequences in the Ombonde-Aap basin (Fig. 1, location 10).

There is evidence from landforms, sediments and soils that the Mid-Holocene more humid conditions ended after 4 ka in parts of the Kalahari (e.g. HEINE 1982; 1990; BEAUMONT et al. 1984; DEACON & LANCASTER 1988). At Garubib Farm (Huab River; Fig. 1, location 9) desert loess was deposited. From shelf sediments it is to be assumed that subsequent climatic oscillations deviated only little from present conditions (e.g. GINGELE 1997).

5 Conclusion and links to global-scale climatic events

The data compiled reveal the climatic trends after the LGM in southwestern Africa. The intensity of the Benguela upwelling system controlling the Namib Desert climate changed inversely to the monsoonal influence during the Younger Pleistocene as indicated by marine sediments off Namibia. SUMMERHAYS ET AL. (1995) stated that the Benguela cold water upwelling was more intensive during Intergla-

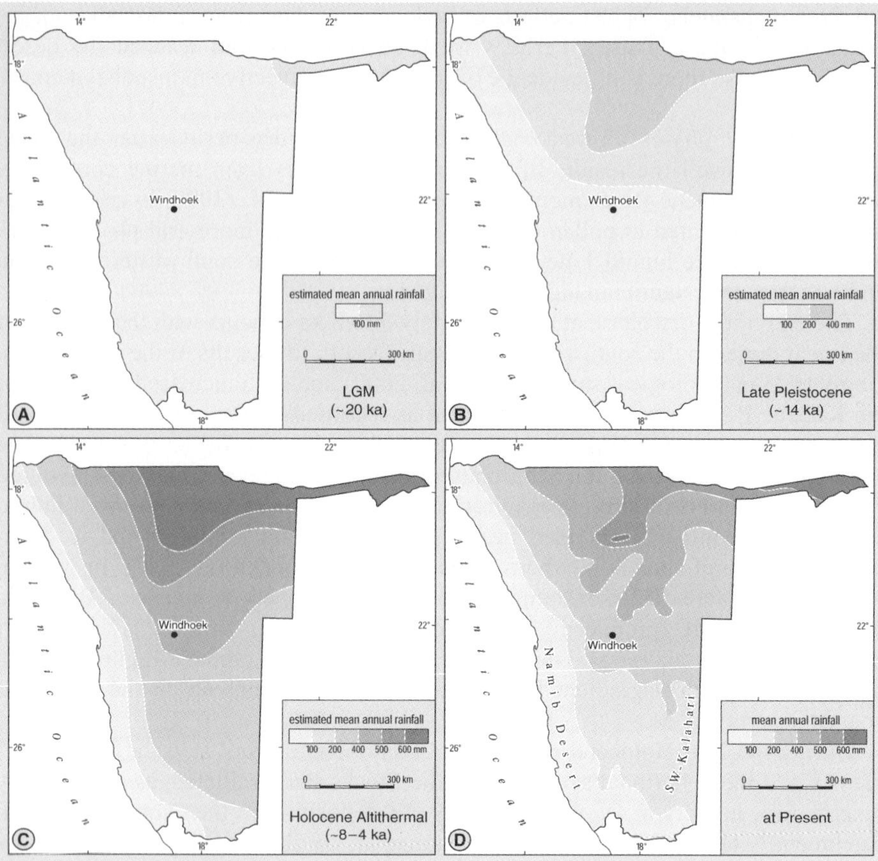

Fig. 5. Sketches to illustrate the transition from arid environments during the LGM (A) to semiarid savanna environments induced by moister conditions advancing from NE to SW. At 14 ka the Teufelsbach sandfield north of Windhoek shifted for the last time, indicating that in northeastern Namibia savanna environments were already established in the Latest Pleistocene and persisted up to the present (B). More humid ecosystems subsequently advanced further to the south and west due to intensified monsoonal influence which reached its maximum at 9 – 8 ka (C), as indicated by dune fixation even in the arid core area of the Kalahari basin and by dissection of fine grained fills east of the Namib Desert margin. After approximately 4 ka aridification led to conditions which prevailed – modified by weak fluctuations- up to the present (D).

cial periods, thereby producing hyperarid conditions, and somewhat reduced during Glacial conditions. Therefore it is possible to exclude aridification effects from the marine system on the climate in the hinterland of the coastal desert during oxygen stage 2.

Between 20 ka to 8 ka moister conditions gradually advanced from northeast to southwest, crossing the Kalahari basin as far as its dry core in southern Namibia

(Fig. 5). At approximately 14 ka the Windhoek highlands changed from desert with active dune formation to a semiarid savanna-type ecosystem. After c. 8 ka the monsoonal influence was so strong that it not only affected the driest part of the Kalahari basin (southeastern Namibia) but also reached the present eastern margin of the Namib Desert. Rainfall intensities increased and fluvial erosion and dissection of fine-grained Pleistocene deposits set in. River silt terraces in different headwater areas of the ephemeral rivers provide evidence for this geomorphodynamic change. In southern Namibia desert environments changed at least to thornbush savannas, and it seems likely that in northern Namibia extended savanna woodlands were established. In the transition zone from the northern Namib Desert to the savanna environments of the Namibian highlands the sensible pedogeomorphic system reveals that the Late Pleistocene-Early Holocene transition from arid to semiarid ecosystems in southwestern Africa was not a continuous process but characterized by intercalated weak hygric fluctuations.

New climatological studies suggest that links exist between summer rainfall intensity over southern Africa and Indian Ocean sea-surface temperatures (LANDMAN & MASON 1999) and/or Walker cell anomalies over the tropical Atlantic (JURY & ENGERT 1999). It seems likely that Late Quaternary environmental modifications in southwestern Africa were signs of hemispheric or global climatic change as they are at present. In several other regions in the southern Hemisphere the moister Mid-Holocene conditions are confirmed too, e.g. in large parts of the central and northern Kalahari basin (PARTRIDGE 1997), by proxy data from Lake Tritrivakely, Madagascar (WILLIAMSON et al. 1998) and from Lake Silvana, SE-Brazil (RODRIGUES-FILHO & MÜLLER 1999). The environmental change in southwestern Africa agrees with the recovery of the tropical rainforests during the Early Holocene (eg. READING et al. 1995).

Environmental change in southwestern Africa coincides with Early Holocene climatic instabilities as indicated by ice-core proxies. The reasons are not clear. Perhaps the change reflects the orbitally induced increase in summer insolation over the southern Hemisphere (COMAP MEMBERS 1988). Paleoclimatic records from Greenland, Antarctica, southern Germany and a diatom record from Lake Victoria, East Africa, reveal that the global atmospheric circulation did not change gradually but abruptly at the Early to Mid-Holocene transition to fully postglacial conditions (ALLEY et al. 1997, STAGER & MAYEWSKI 1997; VON GRAFENSTEIN et al. 1998).

This could have been supported by relatively high temperature gradients between the northern Mid-Latitudes and the Tropics. Increased meltwater inputs during Younger Dryas time and at 8 ka caused a substantial weakening of the thermohaline circulation, with a cooling of the surface and deep waters in the northern Atlantic Ocean and a warming of up to 3-4°C in the equatorial Atlantic Ocean down to intermediate water depths. The thermal equator and the associated low-pressure trough were displaced southward relative to their present position during deglaciation, reaching extremes at 11-10 ka, and for the last time at 8 ka (MULITZA & RÜHLEMANN 2000). This model may explain the rapidly increased influx of moist air from the inner Tropics into southwestern Africa since approximately 8 ka.

6 Acknowledgement

We would like to thank Professor U. RADTKE (Geographisches Institut der Universität Köln) for the optical dating of sand from the Teufelsbach sandfield (Windhoek Highlands) and Professor L. ZÖLLER and Dr. B. MAUZ (Geographisches Institut der Universität Bonn) for the dating of fluvial deposits and desert loess (samples HDS 646-649 and D18-25, Table 1-4). We are grateful to the Deutsche Forschungsgemeinschaft (Bonn) for the financial support of our projects in different parts of Namibia.

7 References

ALLEY R. B., MAYEWSKI, P.A., SOWERS, T., STUIVER, M., TAYLOR, K. C. and CLARK, P. U. (1997). Holocene climatic instability: A prominent, widespread event 8200yr ago. Geology **25**: 483-486.

BEAUMONT, P. B., VAN ZINDEREN-BAKKER, E. M. and VOGEL, J. C. (1984). Environmental changes since 32.000 BP at Kathu Pan, Northern cape. In: Vogel, J. C. (Ed.). Late Cainozoic Palaeoclimates of the Southern Hemisphere. Rotterdam: 329-338.

BESLER, H. (1972). Klimaverhältnisse und klimatische Zonierung der zentralen Namib (Südwestafrika). Stuttgarter Geographische Studien **82**: pp. 209.

BESLER, H. (1991). Der Namib Erg: Älteste Wüste oder älteste Dünen? Geomethodica **16**: 93-122.

BESLER, H. (1992). Geomorphologie der ariden Gebiete. Darmstadt: pp.189.

BLÜMEL, W.D. (1981): Pedologische und geomorphologische Aspekte der Kalkkrustenbildung in Südwestafrika und Südostspanien. Karlsruher Geographische Hefte **10**: pp. 228

BLÜMEL, W.D. (1982). Calcretes in Namibia and SE-Spain – Relations to substratum, soil formation and geomorphic factors. Catena Supplement **1**: 67-82.

BLÜMEL, W.D., EITEL, B. & LANG, A., (1998). Dunes in southeastern Namibia: Evidence for Holocene environmental changes in the southwestern Kalahari based on thermoluminescence data. Palaeogeography, Palaeoclimatology, Palaeoecology **138**: 139-149.

BLÜMEL, W. D., HÜSER, K. & EITEL, B. (2000): Uniab-Schwemmfächer und Skelettküsten-Erg: Zusammenspiel von äolischer und fluvialer Dynamik in der nördlichen Namib. Regensburger Geographische Schriften **33**: 37-56.

BRUNOTTE, E. & SPÖNEMANN, J. (1997). Die kontinentale Randabdachung Nordwestnamibias: eine morphotektonische Untersuchung. Petermanns Geographische Mitteilungen **141**: 3-15.

BUCH, M.W., ROSE, D. & ZÖLLER, L. (1992). A TL-calibrated pedostratigraphy of the western lunette dunes of Etosha Pan / northern Namibia: Palaeoenvironmental implications for the last 140ka. Palaeoecology of Africa **23**: 129-147.

BUCH, M.W. (1996). Geochrono-Geomorphostratigraphie der Etosha-Region, Nord-Namibia. Die Erde **127**: 1-22.

COCKBURN, H. A. P., BROWN, R. W., SUMMERFIELD, M. A. & SEIDL, M. A. (2000): Quantifying passive margin denudation and landscape development using a combined fission-track thermochronology and cosmogenic isotope analysis approach. Earth and Planetary Science Letters **179**: 429-435.

COHMAP MEMBERS (1988). Climatic changes of the last 18,000 years: observations and model simulations. Science **241**: 1043-1052.

DEACON, J. & LANCASTER, N. (1988). Late Quaternary Palaeoenvironments of Southern Africa. Oxford Scientific Publications, Oxford: pp. 225.

EITEL, B. (1993): Kalkkrustengenerationen in Namibia: Carbonatherkunft und genetische Beziehungen. Die Erde **124**: 85-104

EITEL, B., (1994). Paläoklimaforschung: Pedogener Palygorskit als Leitmineral? Die Erde **125**: 171-179.

EITEL, B., (1996). Großräumige Epirogenese und Bruchtektonik östlich der Großen Randstufe in Namibia: Überblick und mögliche Beziehungen zu neotektonischen Leitlinien im südlichen Afrika. Die Erde **127**: 113-126.

EITEL, B. & BLÜMEL, W.D. (1997). Pans and dunes in the southwestern Kalahari (Namibia): Geomorphology and evidence for Quaternary paleoclimates. Zeitschrift für Geomorphologie N. F. Suppl.-Bd. **111**: 73-95.

EITEL, B. & ZÖLLER, L. (1996). Soils and sediments in the basin of Dieprivier-Uitskot (Khorixas District / Namibia): age, geomorphic and sedimentological investigation, palaeoclimatic interpretation. Palaeoecology of Africa **24**: 159-172.

EITEL, B., BLÜMEL, W. D. & HÜSER, K. (1999). River silt terraces at the eastern margin of the Namib Desert (NW-Namibia): Genesis and paleoclimatic evidence. Zbl. Geol.Paläont. Teil I, 1998(5-6): 243-254.

EITEL, B. (2000). Different amounts of pedogenic palygorskite in South West African Cenozoic calcretes: Geomorphological, pelaeoclimatical and methodological implications. Zeitschrift für Geomorphologie N. F. Suppl.-Bd. **121**: 139-149.

EITEL, B., BLÜMEL, W.D., HÜSER, K. AND MAUZ, B. (2001). Dust and loessic alluvial deposits in northwestern Namibia (Damaraland, Kaokoveld): sedimentology and palaeoclimatic evidence based on luminescence data. Quaternary International **76/77**: 57-65.

EITEL, B. & EBERLE, J. (2001): Kastanozems in the Otjiwarongo region (Namibia): Pedogenesis, associated soils, evidence for landscape degradation. Erdkunde **55/1**: 21-31.

EITEL, B., EBERLE, J. & KUHN, R. (2002). Holocene environmental change in the Otjiwarongo thornbush savanna (Northern Namibia): Evidence from soils and sediments. Catena, 47:43-62

EITEL, B., BLÜMEL, W.D., HÜSER, K. ZÖLLER, L. & MAUZ, B. (in prep.): Desert loess in the upper Huab River valley, northwestern Namibia.

GEYH, M.A. & EITEL, B. (1998): Radiometric dating of young and old calcrete. Radiocarbon **40**: 795-802.

GINGELE, F. X. (1997). Holocene climatic optimum in Southwest Africa - evidence from the marine clay mineral record. Palaeogeography, Palaeoclimatology, Palaeoecology **122**: 77-87.

GOUDIE, A. S. (1972). Climate, weathering, crust formation dunes and fluvial features of Central Namib Desert, near Gobabeb, South West Africa. Madoqua **II, 1**: 15-32.

GOUDIE, A. S. & WELLS, G. L. (1995). The nature, distribution and formation of pans in arid zones. Earth Science Reviews **38**: 1-69.

HEINE, K. (1981): Aride und pluviale Bedingungen während der letzten Kaltzeit in der Südwest-Kalahari (südliches Afrika). Z. Geomorph. N. F., Suppl.-Bd. **38**: 1-37

HEINE, K. (1990): Some observations concerning the age of the dunes in the western Kalahari and palaeoclimatic implications. Palaeoecology of Africa **21**: 161-178.

HEINE, K. (1998): Climate change over the past 135,000 years in the Namib Desert (Namibia) derived from proxy data. Palaeoecology of Africa **25**: 171-198.

HÜSER, K., BLÜMEL W. D. & EITEL, B. (1998): Geomorphologische Untersuchungen an Rivierterrassen im Mündungsbereich des Uniab (Skelettküste/NW-Namibia). Zbl. Geol. Paläont. Teil I: **1997,1/2**: 1-21.

HÜSER, K., BLÜMEL, W. D. & EITEL, B., (1998): Landschafts- und Klimageschichte in Südwestafrika. Geographische Rundschau, **1998(4)**: 238-244.

JURY, M. R. & ENGERT, S. (1999): Teleconnections modulating inter-annual climate variability over northern Namibia. International Journal of Climatology **19/13**: 1459-1475.

LANCASTER, N. (1981): Palaeoenvironmental implication of fixed dune systems in southern Africa. Palaeogeography, Palaeoclimatology, Palaeoecology **33**: 327-346.

LANCASTER, N. (1989): Late Quaternary Palaeoenvironments of the southwestern Kalahari. Palaeogeography, Palaeoclimatology, Palaeoecology **70**: 367-376.

LANCASTER, N. (1988): Development of linear dunes in the southwestern Kalahari. J. Arid Environments **14**: 233-244.

LANDMAN, W. A. & MASON, S. J. (1999): Change in the association between Indian Ocean sea-surface temperatures and summer rainfall over South Africa and Namibia. International Journal of Climatology **19/13**: 1477-1492.

LINDEQUE, M. & ARCHIBALD, T.J. (1991): Seasonal wetlands in Owambo and Etosha National Park. Madoqua **17(2)**: 129-133.

LIVINGSTONE, I. & THOMAS, D.S.G. (1993): Modes of linear dune activity and their palaeoenvironmental significance: an evaluation with reference to southern African examples. In: PYE, K. (ed.). The dynamics and environmental context of aeolian sedimentary systems. Geological Society Special Publication **72**: 91-102.

MARKER, M.E. & MÜLLER, D. (1978): Relict vlei silts of the middle Kuiseb River valley, South West Africa. Madoqua **11/2**: 151-162.

MARTIN, H. (1969): Paläomorphologische Formelemente in den Landschaften Südwestafrikas. Geologische Rundschau **58**: 121-128.

MULITZA, S. AND RÜHLEMANN, C. (2000): African monsoonal precipitation by interhemispheric temperature gradients. Quaternary Research **53/2**: 270-274.

NETTERBERG, F. (1969). Ages of calcretes in southern Africa. S. Afr. arch. Bull. **24**: 88-92.

NEWELL, R.E., GOULD-STEWART, S. & CHUNG, J.C. (1981): Possible interpretation of palaeoclimatic reconstructions for 18,000 BP in the region 60° N to 60° S, 60° W to 100° E. Palaeoecology of Africa **13**: 1-19

OLLIER, C.D. (1977): Outline geological and geomorphic history of the Central Namib Desert. Madoqua **10/3**: 207-212.

PARTRIDGE, T.C., (1997): Cainozoic environmental change in southern Africa, with special emphasis on the last 200 000 years. Progress in Physical Geography **21(1)**: 3- 22.

PARTRIDGE, T. C., SCOTT, L. & HAMILTON, J. E. (1999): Synthetic reconstruction of Southern Africa environments during the Last Glacial Maximum (21-16 kyr) and the Holocene Altitherma (8-6 kyr). Quaternary International **57/58**: 207-214.

PASSARGE, S. (1904): Die Kalahari. Versuch einer physisch-geographischen Darstellung der Sandfelder des südafrikanischen Beckens. Berlin: pp. 822.

READING, A. J., THOMPSON, R. D. & MILLINGTON, A. C. (1995): Humid tropical environments. Oxford, Cambridge/Mass.: pp. 429.

RODRIGUES-FILHO, S. & MÜLLER, G. (1999): A Holocene sedimentary record from Lake Silvana, SE Brazil. Lecture Notes in Earth Sciences **88**: pp. 96.

RUST, U. & WIENEKE, F. (1978): A reinvestigation of some aspects of the evolution of the Kuiseb River valley up-stream of Gobabeb, South West Africa. Madoqua **12(3)**: 163-173.

RUST, U. (1984): Geomorphic evidence of Quaternary environmental changes in Etosha, South West Africa/Namibia. In: VOGEL, J.C. (ed.): Late Cainozoic palaeoclimates of the southern hemisphere. Rotterdam: pp. 279-286.

RUST, U., (1987): Geomorphologische Forschungen im südwestafrikanischen Kaokoveld zum angeblichen vollariden quartären Kernraum der Namibwüste. Erdkunde **41**: 118-133.

RUST, U. & VOGEL, J.C., (1988): Late Quaternary environmental changes in the Namib Desert as evidenced by fluvial Landforms. Palaeoecology of Africa **19**: 127-137.

RUST, U., (1989a): Grundsätzliches über Flußterrassen als paläoklimatische Zeugen in der südwestafrikanischen Namibwüste. Palaeoecology of Africa **20**: 119-132.

RUST, U., (1989b): (Paläo)- Klima und Relief: Das Reliefgefüge der südwestafrikanischen Namibwüste (Kunene bis 27° s. Br.). Münchner Geographische Abhandlungen **7**: pp. 158 .

RUST, U. & VOGEL, J.C. (1990): Ein in der kleinen Eiszeit (Little Ice Age) begrabener Wald in der nördlichen Namib. Berliner Geographische Studien **30**: 15-34.

RUST, U. (1999): River-end deposits along the Hoanib River, northern Namib: Archives of Late Holocene climatic variation on a subregional scale. South African Journal of Science **95(4)**: 205-208.

SARNTHEIN, M. (1978): Sand deserts during glacial maximum and climatic optimum. Nature **272**: 396-398.

SCHEEPERS, A. C. T. & RUST, I. C. (1999): The Uniab River fan: an unusual alluvial
fan on the hyper-arid Skeleton Coast, Namibia. In: Miller, A. I. & Gupta, A. : Varieties of fluvial form. Wiley, Chichester: 273-294.

SHAW, P.A. & COOKE, H. J. (1986): Geomorphic evidence for Late Quaternary palaeoclimates of the middle Kalahari of western Botswana. Catena **13**: 349-359.

SHAW, P. A. & THOMAS, D. S. G. (1988): Lake Caprivi: a late Quaternary link between the Zambezi and middle Kalahari drainage systems. Zeitschrift für Geomorphologie 32: 329-337.

SHAW, P.A. & THOMAS, D.S.G. (1996): The Quaternary palaeoenvironmental history of the Kalahari, Southern Africa. J. Arid Environments 32: 9-22.

SHI, N., DUPONT, L. M., BEUG, H.-J. & SCHNEIDER, R. (2000): Correlation between vegetation in Southwestern Africa and oceanic upwelling in the past 21,000 years. Quaternary Research **54**: 72-80.

SMITH, R. M. H., MASON, T. R. & WARD, J. D. (1993): Flashflood sediments and ichnofazies of the Late Pleistocene Homeb Silts, Kuiseb River, Namibia. Sedimentary Geology **85**: 579-599.

STAGER, J. C. & MAYEWSKI, P. A. (1997): Abrupt early to mid-Holocene climatic transition registered at the equator and the poles. Science **276**: 1834-1836.

STOKES, S., THOMAS, D. S .G. & WASHINGTON, R., (1997). Multiple episodes of aridity in southern Africa since the last interglacial period. Nature **388**: 154-158.

SUMMERHAYS, C. P., KROON, D., ROSELL-MELÉ, A., JORDAN, J. D., SCHRADER, H., HEARN, R., VILLANUEVA, J., GRIMALT, J. D. & EGLINGTON, G. (1995): Variability in the Benguela Current upwelling system over the past 70,000 years. Progress in Oceanography **35**: 207-251

THOMAS, D. S. G. (1987): Discrimination of depositional environments, using sedimentary charcteristics, in the Mega Kalahari, central southern Africa. In: FORSTICK, L. E. & REID, I. (eds.). Desert sediments, Ancient and Modern. Oxford: pp. 293-306.

Thomas, D. S.G. & Shaw, P.A. (1991): The Kalahari environment. Cambridge, pp. 284.

THOMAS, D.S.G., O'CONNOR, P.W., BATEMAN, M.D., SHAW, P.A., STOKES, S. & NASH, D.J. (2000). Dune activity as a record of late Quaternary aridity in the Northern Kalahari: new evidence from northern Namibia interpreted in the context of regional arid and humid chronologies. Palaeogeography, Palaeoclimatology, Palaeoecology **156**: 243-259

TYSON, P.D. (1986): Climatic change and variability in Southern Africa. Oxford, 220 pp.

VAN DER WATEREN, F. M. & DUNAI,T. J. (2001): Late Neogene passive continental margin denudation history - cosmogenic isotope mesurements from the central namib desert. Global and Planetary Change **628**: (in press)

VOGEL, J.C. (1982): The Age of the Kuiseb River Silt Terrace at Homeb. Palaeoecology of Africa **15**: 201-209.

VON GRAFENSTEIN, U., ERLENKEUSER, H., MÜLLER, J., JONZEL, J. & JOHNSON, S. (1998): The cold event 8200 years ago documented in oxygen isotope records of precipitation in Europe and Greenland. Climate Dynamics **14**: 73-81.

WARD, J.D., SEELY, M.K. & LANCASTER, N. (1983): On the antiquity of the Namib. S. Afr. J. Sci. **79**: 175-183

WATTS, N.L. (1980): Quaternary pedogenic calcretes from the Kalahari (southern Africa): mineralogy, genesis and diagenesis. Sedimentology **27**: 661-686.

WILLIAMS, G. J. (1986): A preliminary Landsat interpretation of the relict land forms of western Zambia. In: WILLIAMS, G. J. & WOOD, A. P. (Eds): Geographical Perspectives on Development in Southern Africa. Commonwealth Geographical Bureau, James Cook University, Queensland: 23-33.

WILLIAMSON, D., JELINOWSKA, A., KISSEL, C., TUCHOLKA, P., GIBERT, E., GASSE, F., MASSAULT, M., TAIEB, M., VAN CAMPO, E. & WIECKOWSKI, K. (1998): Mineral-magnetic proxies of erosion/oxidation cycles in tropical maar-lake sediments (Lake Tritrivakely, Madagascar): paleoenvironmental implications. Earth and Planetary Science Letters **155**: 205-219.

Aeolian sedimentation in arid and semi-arid environments of Western Mongolia

Jörg Grunert[a] & Frank Lehmkuhl[b]

[a] Geographisches Institut der Universität Mainz, Becherweg 21, D-55128 Mainz, Germany

[b] Geographisches Institut der RWTH Aachen, D-52056 Aachen, Germany

Summary

Research on aeolian sediments in Mongolia shows two main cycles of aeolian sedimentation: first the accumulation of major sand fields neighbouring the eastern bank of rivers and lakes, and second the distribution of loess-like sediments on the mountain slopes. The first is resulting from strong westerly winds, being more strength especially during the glacial periods. The latter is resulting in the erosion and accumulation of silt in this region in more humid periods in Interstadial stages and at the end of glacial periods. Both cycles are described on the case study area of the Uvs Nuur Basin in Western Mongolia.

1 Introduction

This paper presents details on sand dunes and on loess-like sediments in the area of the Mongolian Altai in western Mongolia and its Late Quaternary evolution. Concerning aeolian processes in Central Asia, there is a considerable literature on the Quaternary loess in the deserts, especially within China. In addition, the loess sequences in the Chinese Loess Plateau are the best-known and most intensively studied within China (e.g. AN et al., 1991; HOVAN et al., 1989; LIU et al., 1985, 1986), which provides long-time records for almost the whole Quaternary (e.g. DING et al., 1992).

However, although there is a remarkable extent of aeolian material in Mongolia, only a few papers have focused on them, especially in the western literature. First MURZAEV (1954) published the distribution of the major sand areas in Mongolia. More recent studies on sand dunes in Western Mongolia, especially the Uvs Nuur Basin, are presented by DASH (1999) and GRUNERT et al. (1998, 1999, 2000). Russian and Mongolian scientists mentioned some areas covered with loess and loess-like sediments only in the northern part of central Mongolia, in the vicinity of the rivers Orchon and Selenga (DORDSCHGOTOV, 1992). The supposed area with loess and the distribution of sand is shown in Fig. 1. FENG et al. (1998) and FENG (2001) described recently loess sequences in the Buregkhanga area in Central Mongolia (104°E, 48°N). In this region, loess-palaeosol sequences provide sedimentological evidence for dominant aeolian activity between 40 and 30 ka, and colluvial activity from 30 ka until 24 ka (FENG et al., 1998, 2001). This suggests more humid conditions in the latter (30-24 ka) with respect to the earlier (40-30 ka) times. LEHMKUHL (1997b) reported on loess and loess-like sediments in the mountains of Central Asia, mainly the Tibetan Plateau. First results on the Turgen-Kharkhiraa Mountains, the northernmost part of the Mongolian Altai (Fig. 1), were presented in 1999 and 2000 (LEHMKUHL, 1999a; LEHMKUHL et al., 2000).

Recent comprehensive studies on Late Quaternary lake level fluctuations in Central Asia and Tibet are given e.g. by FANG (1991), QIN and YU (1998), TARASOV el al. (1996), and TARASOV and HARRISON (1998). Concerning lake level fluctuations in the Uvs Nuur Basin, NAUMANN (1999), NAUMANN and WALTHER (2000) provided dates from geomorphological and sedimentological research from the Bayan Nuur, and WALTHER (1999) from the Uvs Nuur, respectively. A review and discussion of lake level history, the fluctuation of mountain glaciers, and other Late Quaternary palaeoclimatic implications from Central Asia are presented by FRENZEL (1994), and LEHMKUHL and HASELEIN (2000). KLINGE (2001) and LEHMKUHL (1998) focused on the modern and Pleistocene glaciations of western Mongolia.

The major sand areas of Mongolia and northern China are shown in Fig. 1. Dunes are concentrated on three areas: The large dunefields of Jungaria south of the Mongolian Altai and of the Badain Jaran Desert in the southern Gobi, very close to the Chinese / Mongolian border and, in the far west of Mongolia, the smaller dunefields between Mongol Altai and Khangay. Numerous but small dune fields are located in the basin and range area of southern Mongolia, east of the Gobi Altai.

The origin of dunesands in the southern part of the former Sovjet Union has been discussed by BERG (1958). In general, they represent reworked fluvial sediments of large rivers, which are originating in high mountain ranges. This model

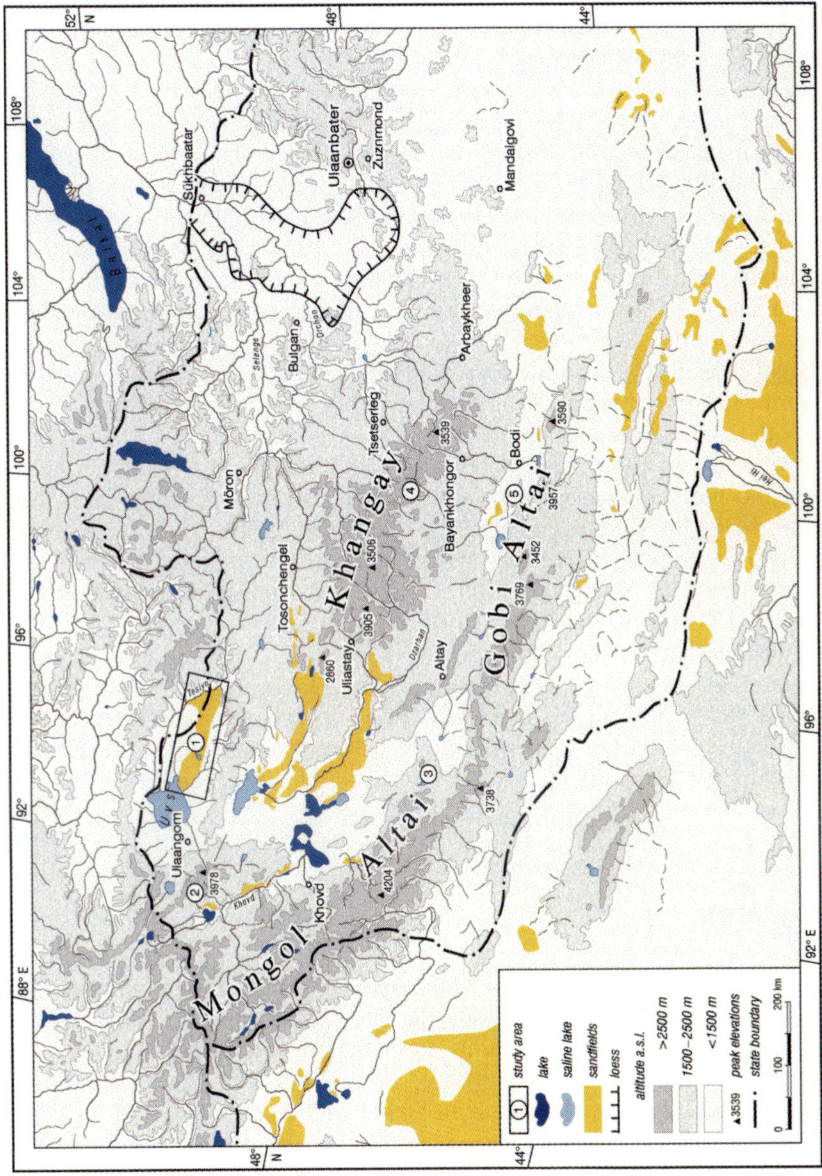

Fig. 1. Topographic map of Western Mongolia showing the widespread sandfields and the different study areas (LEHMKUHL)

can be adapted to Mongolia and Western China. In case of the Jungaria dune fields, there is a close connection with the rivers from the high mountain range (5,500 m) in the south. The numerous barkhanes and barkhanoid dunes indicate prevailing WNW – winds. Due to these wind systems, they are limited to the inner basin in elevations below 600 m a.s.l..

The same situation can be observed in Badain Jaran Desert (JÄKEL, 1995, 1996). The Hei He River west of this desert is regarded to be the main sand source. The main River is originating in the Qilian Shan, the north-eastern fringe of the Tibetan Plateau (6,300 m) and terminates in the Gaxun and Sogun Nuur Lake depression (820 m). The Late Quaternary lake level fluctuations in this area have been studied by WÜNNEMANN (1999) and HOFMANN (1999). The giant sand sea of Badain Jaran can be interpreted as reworked fluvial sediments, which have been accumulated after a short-distance transportation by prevailing westerly winds.

No connection can be seen between the Badain Jaran and the small dunefields in southern Mongolia. Fieldwork in the years 1994 and 2000 revealed their connection with alluvial fans and fanglomerates of local river systems. The mountain ranges of the eastern Gobi Altai system reach elevations up to 2,200 – 2,500 m. However, the northern part of the Gobi Altai rises up to 3,957 m (Ikh Bogd Mountain), being the southern fringe of the so-called Valley of the Gobi Lakes. This is a graben zone between the southern slope of the Khangay and the Gobi-Altai. Rivers from the Khangay (3,500 – 3,700 m) with high water volume feed several lakes (LEHMKUHL and LANG, 2001; MURZAEV, 1954).

The dunefields of western Mongolia are of medium size and obviously connected with rivers and lakes. These represent endorheic depressions, the bottom of which lies 1,130 m a.s.l. in the south and 760 m a.s.l. in the north. The fact that dunefields are climbing the western flanc of Khangay Mountains up to 2,300 m a.s.l. may indicate a very effective WNW- wind system. This could be interpreted as a leeward effect of the Russian Altai.

2 Study area

The study area is situated in Western Mongolia comprising the high mountains of Mongolian Altai, southern Khangay and the large endorheic depression between the two mountain systems (Fig. 1). In its northern part, the so-called Valley of the Great Lakes (FLORENSOV and KORZHNEV, 1982; MURZAEV, 1954), three local dunefields of about 200 km length and 30-50 km width have developed: the Mongol Els, the Borkhar Els and, very close to the Russian border and east of the huge lake Uvs Nuur, the Böörög Deliyn Els (DASH, 1999). Despite its location at the same latitude as Central Europe (50°N), this represents the northernmost dunefield of the Central Asian arid belt. This paper focuses mainly on the Uvs Nuur Basin and the surrounding mountain ranges, especially the high massif at its southwestern rim, the Turgen-Kharkhiraa (3,978 m). The investigations have been part of a German-Mongolian research project (1994-2000).

The endorheic Uvs Nuur Basin stretches about 300 km from west to east and 150 km from north to south covering an area of some 45,000 km^2. High mountains of 3,000 m altitude are bordering the basin in the north (Tannu Ola), in the south (Khan Khökhiyn Nuruu, 2,900 m), and in the east (northern Khangay, 2,100 m). The locally important massif of Turgen-Kharkhiraa forms the south-western border. These mountains are the northernmost part of the Mongolian Altai. The highest summit reaches 3,978m a.s.l., and it covers an area of approximately 5,700 km^2. Palaeozoic granites and gneiss build up its central part, whereas metamorphic and sedimentary rocks form the margins (LEHMKUHL, 1999a). The adjacent basins are covered by thick layers of fanglomerates, dune sands and lacustrine sediments (DEVJATKIN and MURZAEV, 1989; WALTHER and NAUMANN, 1997). The whole area is dissected by numerous active faults; one of the most important ones is bordering the northern flanc of Khan Khökhiyn Nuruu Mts. Earthquakes with a magnitude up to 7.5 (R) have been recorded here.

The summits of Turgen-Kharkhiraa Mts. are glaciated. Investigations on the modern glaciers of about 26 km^2 and its extension during the glacial periods of the Pleistocene have been carried out by LEHMKUHL (1999a) and LEHMKUHL et al. (1998, 2000).

Despite the discharge of the glacier-nourished Kharkhiraa River is high (about 10 m^3/s) during summer and its length of about 70 km is remarkable, it does not represent the main drainage system of the Uvs Nuur Basin. This is formed by the river Tesijn Gol, 568 km long and with a total catchment area of 33,350 km^2. During the summer months (1996-1998) its discharge has been estimated up to 100 m^3/s, nourishing the huge Uvs Nuur Lake (3,350 km^2), which represents the deepest point (760 m) in the basin. Masses of silty and sandy sediments have been continuously deposited in the large delta of Tesijn Gol during the Pleistocene.

The dunefield Böörög Deliyn Els is originating at the eastern border of the lake. It stretches about 200 km in ESE direction in adaptation to the prevailing wind system, seasonally strong westerlies (April-May). The width is about 30 km and the estimated average thickness of the sand is at least 30 m. Since the bottom of the Uvs Nuur Basin rises continuously in the same direction up to 1,500 m a.s.l., the dunefield culminates at 1,550 m in its eastern part. It is bordered at this end by the large and incised valley of the Tesijn Gol (1,200 m), which is eroding permanently the migrating sand masses. Investigations on the geomorphology and evolution of the dunefield have been carried out by DASH (1999), DASH and TUMURBAATAR (2000) and GRUNERT et al. (1998, 1999). The first description was given by MURZAEV (1954). The steppe vegetation of the Uvs Nuur Basin has been investigated by HILBIG et al. (1999). Owing to higher humidity, the steppe vegetation in the lowland near the Uvs Nuur Lake changes from a semi-desert steppe (*Ephedra sp.*) to a long-grass steppe (*Stipa sp.*) in the eastern, highest part of the dunefield. Here, a very sparse forest of *Larix sibirica* indicates the position of the present-day lower timberline. The mean annual precipitation can be estimated at 200 mm; near the Uvs Nuur lake, the total amount is only 100 mm/a.

The recent continental climatic conditions are characterised by a wide annual range of temperature. For example, the mean temperature at Ulaangom south of the Uvs Nuur Lake is -32.9°C in January and 19.2°C in July with an annual average of -3.7°C (Table 1). Annual rainfall in the basin ranges from 100 mm up to 400 mm (estimated) in the summit area of Turgen-Kharkhiraa. Mean annual rainfall of Ulaangom is 136 mm (1952-1995), but the variability is high, ranging from 62.8 mm (1952) to 225.2 mm (1965). In Baruunturuun, a small town at the southern rim of the basin (1,850 m), the annual precipitation reaches 218 mm (1940-1990). Further information concerning the soil temperatures of Ulaangom and the Turgen-Kharkhiraa Mounatains is provided by LEHMKUHL and KLINGE (2000).

The modern altitudinal belts in the mountains of Mongolia and the investigation area are depending on the general climatic conditions and are described in the literature: for the vegetation (HILBIG et al. 1999), the soils (BATKHISHIG and

Table 1. Monthly mean values of air *temperature* [°C] and precipitation [mm] (average, minimum, maximum) for Ulaangom (UG, 939 m a.s.l.; 1952-1995) and Baruun Turuun (BT, 1940-90, 1850 m a.s.l.; 94°24'E, 49°39'E).

UG	Jan	Feb	Mar	Apr	May	June
Temp.	-32.9	-30.3	-18.9	-0.2	11.2	17.8
Precip.	2.7	2.3	4.0	4.0	7.3	24.2
Min	0	0	0	0	0	0
Max	6.7	6	12	15.4	38.7	87.6
BT						
Temp.	-31.7	-29.9	-18.2	-2.0	9.9	16.4
Precip.	4.3	3.1	7.7	13.2	14.2	27.6

UG	July	Aug	Sep	Oct	Nov	Dec	Year
Temp.	19.1	16.9	10.2	0.2	-11.2	-26.6	-3.7
Precip.	35.4	23.8	15.1	4.7	8.5	4.0	135.7
Min	0	4.2	1.2	0	1.5	0	62.8
Max	92	69.3	63.2	19.9	39.9	14.2	225.2
BT							
Temp.	17.3	15.1	8.7	-1.2	-14.4	-27.0	-4.8
Precip.	55.3	39.8	25.2	12.3	3.2	1.6	218.3

LEHMKUHL 1999; DORDSCHGOTOV, 1992; HAASE, 1978), and the geomorphological processes (RICHTER et al., 1961; KOWALKOWSKI and STARKEL, 1984; LEHMKUHL, 1999a; LEHMKUHL and LANG, 2001, WALTHER, 1998). They are remarkably modified by the different radiation depending on the exposition of slopes. A big contrast exists between north and south facing slopes in the whole area.

3 The cycle of dunesands

The dunefield Böörög Deliyn Els has been geomorphologically investigated during three summer seasons (1996-1998). Different dune types could be mapped representing different periods of formation. It was possible to define dunes of different age by studying their soil and vegetation cover; three major dune generations could be distinguished by GRUNERT et al. (1999, 2000) (Fig. 2).

The oldest one is mostly represented by longitudinal and by giant transversal dunes, respectively. They are covered by dense steppe vegetation (100 %) and a well-developed castanozem (DORDSCHGOTOV, 1992, OPP, 1991). They are fixed today in adaption to a semi-arid climate (150-200 mm/a). In their active phases during the Pleistocene - the last phase is supposed to have been after the LGM (18 –13 ka) - they migrated from west to east cutting off local river systems. The most prominent example is the Baruunturuun River (KLEIN, 2001). New OSL-data given by E. RHODES (Oxford) in 2000 indicate, however, a more recent date of the final activity phase (9.55 ±1.07 ka (OxL-1010); 10.77 ±1.31 ka (OxL-1011); 10.81 ±1.44 ka (OxL-1013), and 11,8 ±0.9 ka, (OxL-1046). It can be interpreted as younger Dryas. Therefore, it is proofed, that the formation of the castanozem covering the old dunes began after the younger Dryas arid phase.

The medium dune generation is predominantly represented by parabolic dunes of an age probably younger than 3,000 y.b.p. Unfortunately, there are no OSL-data available. The dunes are covered by an initial grey soil and sparse steppe vegetation (30-50 %). Parabolic dunes cover more than 50 % of the whole dune area. Normally, they have developed from old dunes due to an aridification of the climate during the younger Holocene and, maybe, they also represent the earliest influence of man-made desertification.

The youngest dune generation is represented by barchans, which are difficult to interpret as climatically induced except by locally strong winds. Desertification processes, however, have formed most of them.

Dunes older than LGM (20 ka) could not be mapped. Therefore, when we started our investigations the dunefield as a whole was supposed to be very young (20-18 ka), according to the arid period between 20 and 13 ka, which has affected Central Asia (FRENZEL, 1994; HOFMANN, 1993; LEHMKUHL and HASELEIN, 2000; PACHUR et al., 1995; WÜNNEMANN et al., 1998). Now a new interpretation based on new OSL-data (E. RHODES) is possible. The samples for dating have been taken

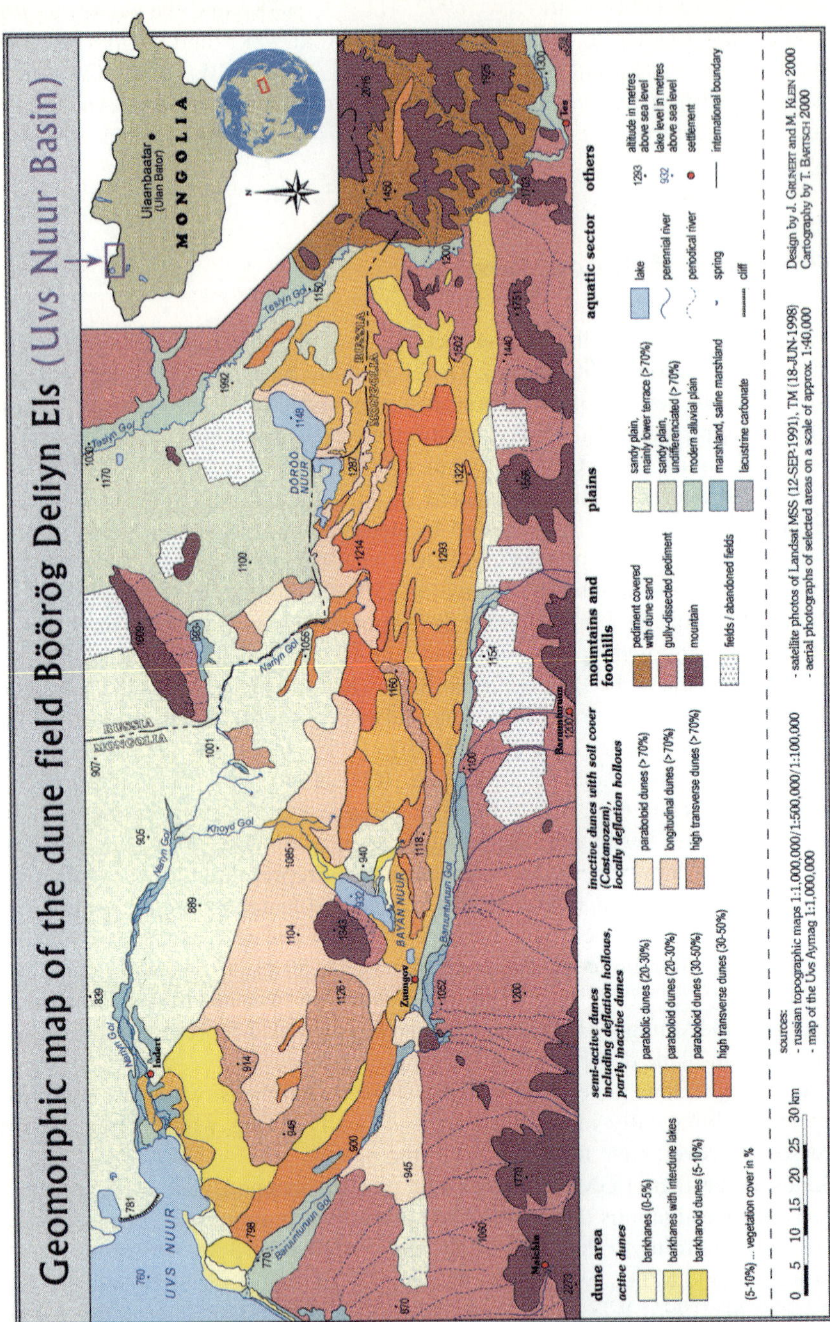

Fig. 2. Geomorphological map of the dunefield Böörög Deliyn Els east of Uvs Nuur Lake showing different dune types according to three generations (GRUNERT).

from a 20 m high dune site in the valley of the river Chusutuin Gol, which is tributary to the lake Bayan Nuur in the centre of the dunefield. The river represents the ancient Baruunturuun Gol, which has been blocked by migrating dunes (KLEIN, 2000). The important site (D, A7) can be divided into three parts: The base consists of pure, unconsolidated dunesand the age of which must be older than 209 ka (mid-Quaternary). The central part of the profile consists of coarse sand and fine gravel deposited by the ancient Baruunturuun Gol (OSL-age of 209 ±26 ka, OxL-1048). This date may be problematic and it is, indeed, unexpectedly high. It can be compared with another OSL-date (181 ±11 ka, OxL-1047) of dunesand belonging to the upper part of the profile. The sample has been taken 2 km upstream at another big dune site.

It can be concluded that at least two important dune formation periods may have existed since the mid-Pleistocene (about 300 ka) indicating an arid climate. It can also be supposed that several smaller arid phases occurred during this long time, indeed, they could not be identified sedimentologically (GRUNERT 2000). In contrast, several lake transgressions during the younger Pleistocene indicate, in alternance, more humid conditions than today (DORFOFEYUK and TARASOV, 1998; NAUMANN, 1999; WALTHER, 1999). However, the present-day relief of the dunefield documented by the geomorphological map (Fig. 2) has been formed completely during the arid phases after the LGM.

Based on these informations, the main problem is now how to explain the dune sand transport over a distance of more than 200 km and to find out the source of the masses of dunesand. It is obvious that strong WNW-winds like today combined with a very sparse vegetation cover were responsible for a very effective sediment transport. The predominance of aeolian processes during arid periods seems to be clear. At the same time it can be supposed that the large Uvs Nuur Lake suffered a regression phase. As it is a shallow lake with a maximum depth of only 25 m (WALTHER, 1999), it is possible that it was completely dried out during periods of maximum aridity in the Quaternary.

Regarding the topographic maps, there is obviously a close connection between the Uvs Nuur Lake and the eastwards adjoining Böörög Deliyn Els. This can be confirmed by granulometric and mineral analyses. The content of carbonate is very high (15-20 %) near the lake; many of the grains could be identified as aragonites. Following the dunefield eastwards, this content diminishes continuously. In samples of fine sand around the lake Bayan Nuur 5-10 % of it has been found, and only 0-3 % in the eastern part of the dunefield. This can be explained as a leaching effect due to an estimated annual precipitation of at least 200 mm.

In contrast to the carbonate content, grain size analyses do not show a clear W-E gradient and, therefore, cannot easily be interpreted. Fine sand is dominating in all samples associated with few silt. This can be demonstrated by three samples, D A18, 10 km east of Uvs Nuur (810m), D A1, 80 km east of Uvs Nuur, very close to Bayan Nuur (1,100 m), and D P6, 170 km east of Uvs Nuur in the highest part of the dunefield (1,500 m, Fig. 3). There is no proof for a growing content of silt eastwards as it could be supposed due to the growing distance from the lake basin. In contrast, the predominance of parabolic dunes in the eastern part of the dunefield far from Uvs Nuur may indicate higher wind velocities. As a result, silt should have been blown out completely and deposited as loess on the flancs of the northern Khangay Mts. Compared with these three dunesand samples the sample of the floodplain of Tesijn Gol near Tes (1250 m) at the eastern border of Uvs Nuur Basin (D P2) is clearly different by its high content of silt. This is a layer interbedded between (fluvial) sand layers granulometry of which resembles that of dunesand (Fig. 3).

Heavy mineral analyses of samples from different parts of the dunefield and, moreover, of samples from fluvial sediments of the Tesijn Gol clearly show similarities with a predominance of instable components (amphibole more than 50 %, clinopyroxene 10-20 % and hypersthene less than 7 %). Granet and epidote range between 2 and 7 %. The stable minerals like zircone and rutile are very rare (0-3 %). Compared with a heavy mineral analysis of a sample of fluvial sand taken from the river Baruunturuun (11), there is a difference: Here the content of hypersthene is about 30 % due to the granites, which are very common in the watershed. The influence of Baruunturuun Gol on the dunefield seems to have been very low at all times. In contrast, the influence of the river Kharkhiraa, which also flows into the Uvs Nuur is estimated to be remarkable. Unfortunately, samples from there are still in preparation (Fig. 4).

As a result, the following model of sediment transport during the younger Pleistocene can be presented: Sandy and silty as well as gravelly sediments are transported continuously by the big river Tesijn Gol and are deposited in the vast plain east of the Uvs Nuur, especially in the large delta. Obviously, during arid periods a critical region was the far eastern part of the dunefield, where sand masses were able to block the river.

During an arid period with a severe regression of the lake the sands and silts could be blown out by strong WNW-winds and transported eastwards where they were deposited continuously. An initial stage of the dunefield was born. This could be mid-Pleistocene in age (about 300 ka, see OSL-dates). In the following period of the younger Pleistocene, we postulate an alternance of arid and semi-arid to semi-humid periods with the consequence of strong fluctuations of the Uvs Nuur lake level (WALTHER, 1999). Correspondingly, there was an alternance between dune formation periods (aeolian activity) and soil formation periods (aeolian stability). At the same time fluvial activity of the rivers was high, like today. During the arid periods the most eastern part of the dunefield was a critical point for the Tesijn Gol, since at this location sand masses are migrating permanently towards the riverbed. But no lacustrine sediments have been found upstream around Tes, which might indicate a dammed lake during an extremely arid period. Therefore it can be concluded that the big river Tesijn Gol transported fluvial sands and gravels as well as eroded dunesand at all times. Correspondingly, it is obvious that the masses of fine sediment deposited in the delta and probably the bottom of Uvs Nuur could be eroded by wind exclusively during arid periods. Therefore, a periodical formation of the dunefield can be postulated (Fig. 5).

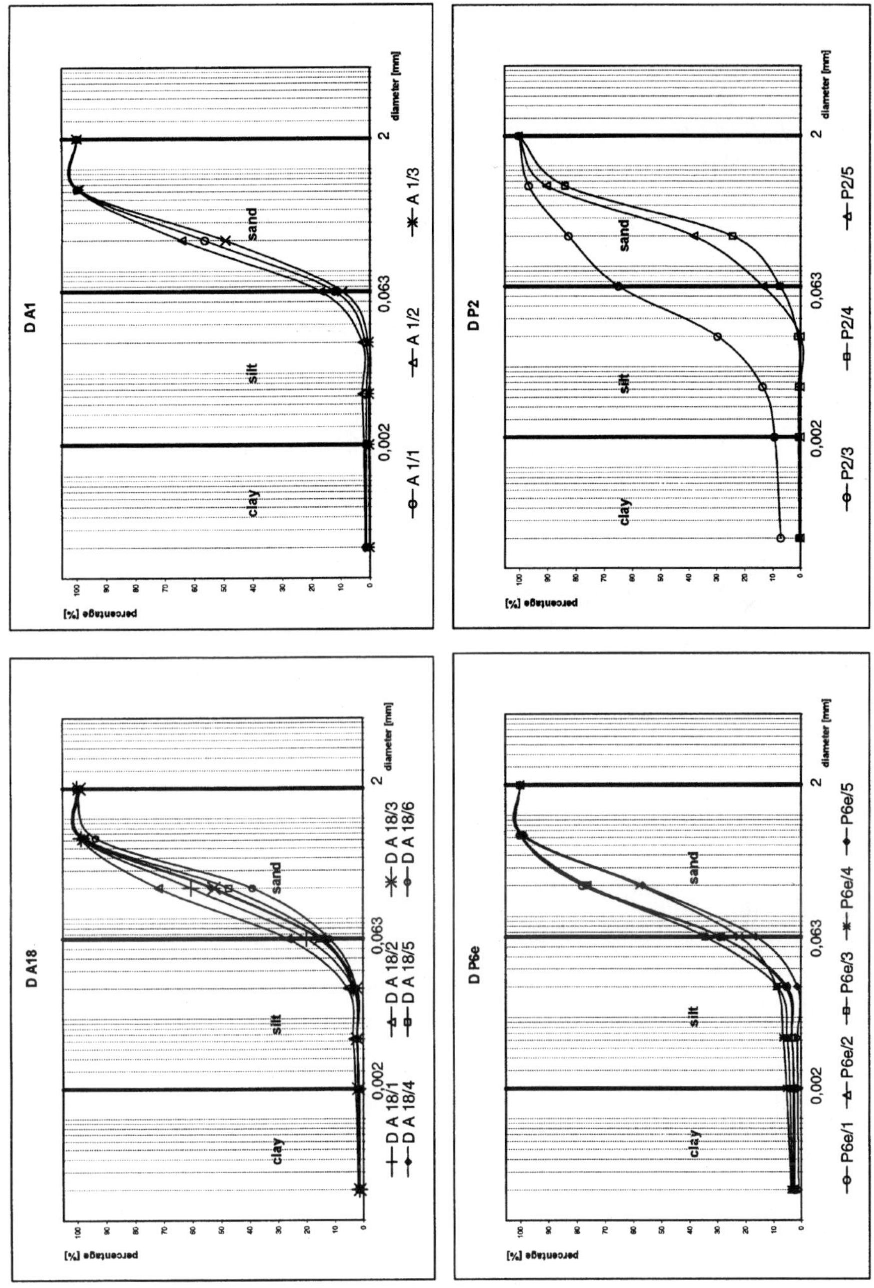

Fig. 3. Summarised grain size diagrams of samples of aeolian and fluvial sand (GRUNERT). The samples correspond with the following heavy mineral samples (see Fig. 4): DA18 = 1, D A1 = 7, D P6e = 13 and D P2 = 15.

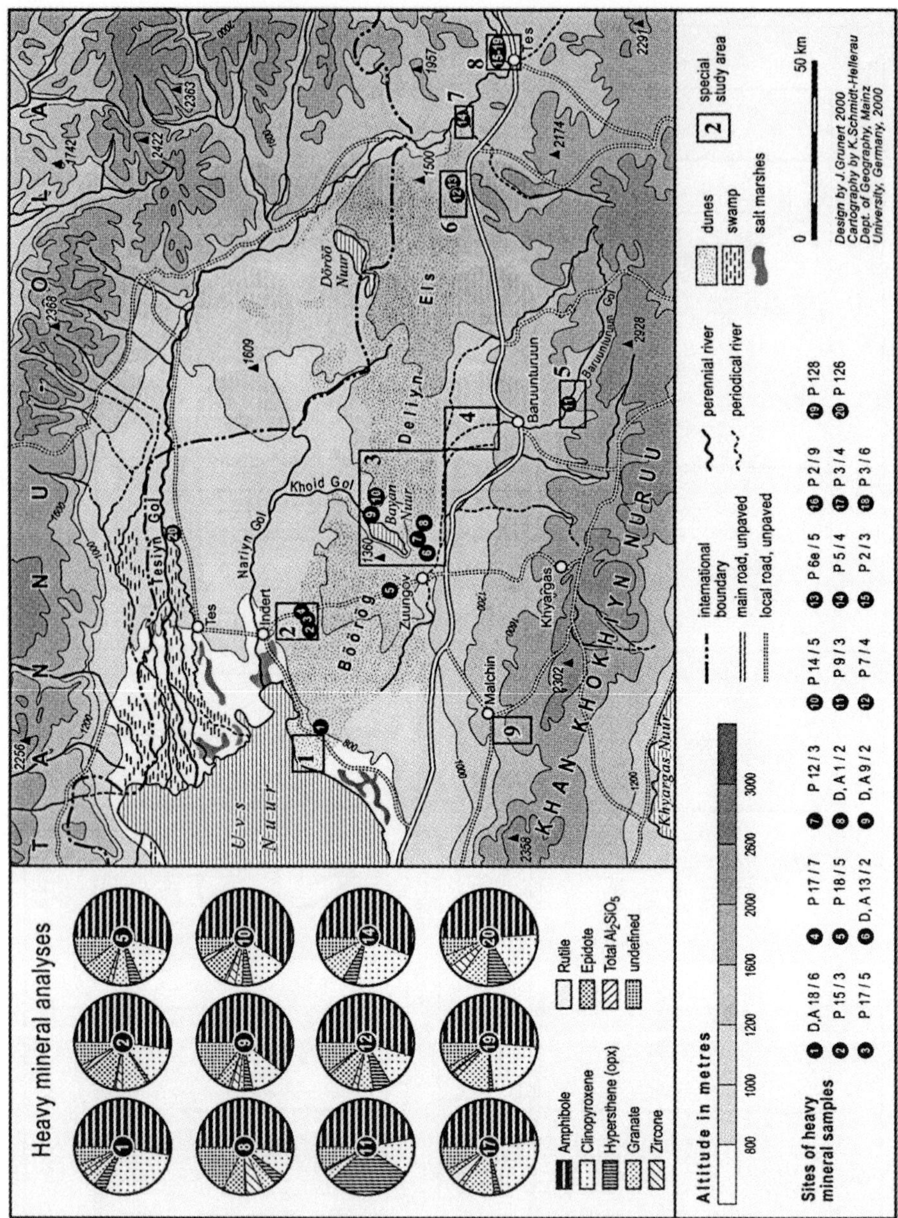

Fig. 4. Sites of heavy mineral samples. Unfortunately, sampling was not possible beyond the Mongolian-Russian border (GRUNERT).

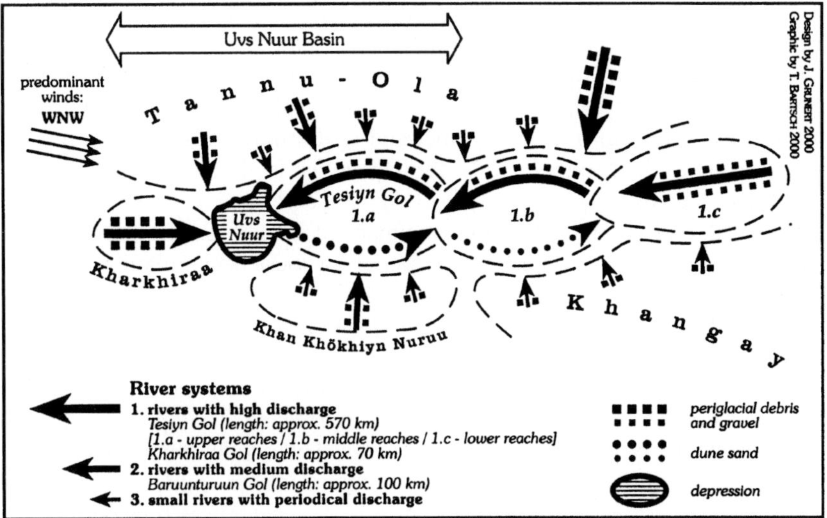

Model of sediment transport

Fig. 5. Model of horizontal transport in the Uvs Nuur Basin during Pleistocene and Holocene in accordance with semi-arid (fluvial activity) and arid (aeolian activity) climatic conditions (GRUNERT).

4 The cycle of loess-like sediments

4.1 Current state of research in Central Asia

First it should be considered, that there are many different definitions of loess in the literature. PÉCSI (1990) listed ten main criteria defining typical (true) loess. In this paper we use a simplified definition as given by PYE (1996: 654): loess is a "...terrestrial clastic sediment, composed predominantly of silt-size particles, which are formed essentially by the accumulation of wind-blown dust".

As mentioned above, references to loess or loess-like sediments in Mongolia are sparse in the literature. HÖVERMANN and HÖVERMANN (1991) postulated for the mountain areas at the southern margin of the deserts of Central Asia that loess or loess-like sediments cover the landscape in the western Kunlun Shan, and that loess occurs elsewhere in the mountain ranges of Xinjing, western China. They noted a loess accumulation zone above 2,500 m a.s.l. In the mountains south of the Qaidam Basin, an eastern extension of the Kunlun system, loess-like sediments occur mainly on north facing slopes between 3,000 and 4,100 m (LEHMKUHL, 1997b). In addition, HÖVERMANN (1987) reported that loess is the dominant surface cover on the east-facing slopes of the Anyêmaqên region (eastern Tibet) at elevations of between 3,500 to 3,900 m, where alpine meadows exist. A sediment cover of aeolian origin is present in several mountain areas of the Tibetan Plateau and Mongolia (LEHMKUHL, 1997b; LEHMKUHL et al., 2000). In the areas above about 3,600 to 4,300 m in eastern Tibet and up to more than 5,000 m in western Tibet, as well as in Mongolia this aeolian cover is dominated by sandy-loess (LEHMKUHL, 1997b: 114).

LEHMKUHL and HASELEIN (2000) summarised studies on the dust deposits of Central Asia and Tibet and presented a model for loess accumulation and fluctuation of lake levels and fanglomerates in the Qaidam Basin. NILSON (1998) and NILSON and LEHMKUHL (2001) summarised the various studies dealing with the Pleistocene variability of flux and accumulation of aeolian dust on the Asian continent, Japan, and the Pacific Ocean. In these comparative studies they show three main patterns in the dust signal for the upper Pleistocene. They propose a tentative model of dust supply for the last 100,000 years. This model elucidates three main aspects. First, the dust supply of the desert regions was maximised during Glacial periods, when most lakes dropped and much rock detritus was transported by episodic floods towards large alluvial fans (pediments and fanglomerates) reaching the basins. Second, a decrease in dust flux observed at the end of the Glacial stages was initially caused by a climatic change towards hyperarid conditions rather than towards more humid conditions. In these periods the runoff from the mountains declined and so did lake levels. Some of the lakes completely dried out and their sands and fine silts were blown out. Sand dunes and sand fields were accumulated on the leeward side of palaeo-lakes and rivers (see Fig. 1). During this phase the dust supply was maintained by the increasingly exposed lacustrine sediments, and possibly by aeolian abrasion in the dune fields. Thereafter conditions became more humid, lake levels rose and the vegetation expanded so that the dust supply was minimised, but the vegetation cover as a main dust trap captured the loess-like sediments in the area. LEHMKUHL and HASELEIN (2000) provide a corresponding example for the first two periods from the Qaidam Basin.

However, for the accumulation of aeolian, loess-like sediments, and the development of aeolian mantles in general the trapping of dust remains the most important process. A denser vegetation cover is commonly regarded as the major operative trap for typical loess (e.g. TSOAR and PYE, 1987). In some regions of Asia, such as high mountain areas and desert margins, the vegetation cover appears to be the dominant determinant of loess deposition. LEHMKUHL (1997b) presents two models of dust accumulation in the mountain areas of Tibet. The air flow is intercepted by a mountain range and the dust is trapped by an increase in the density of vegetation arising from the higher precipitation and the lower temperatures that cause reduction in evapotranspiration at higher altitudes. However, in the mountains of Mongolia as well as in the larger basins of western Tibet the grain size composition of the aeolian mantle is coarser and sandy loess or sand dominates. Especially when the near-surface wind speeds are stronger, or a greater amount of sand is present in the vicinity of the aeolian deposits, accumulation of sandy loess is more likely.

Although there is almost no doubt on the main dust sources of Asia, the precise terrain-types which supply the dust (e.g. PYE and ZHOU, 1989) as well as the areas where the dust particles are produced are not completely understood (e.g. SMALLEY, 1995; DERBYSHIRE et al., 1998). According to recent simulations of WRIGHT et al. (1998) there are several possible processes, which can produce silt-sized particles. However, the relative importance of the individual processes working in the Asian environments is not known.

At present it seems likely that most of the dust is originally formed in mountainous areas with their active periglacial and glacial environments. From here the particles were transported fluvially towards the desert basins. Based on field observations in Tibet and Mongolia the authors share the opinion of many others (e.g. DODONOV, 1991; SMALLEY and KRINSLEY, 1978; HÖVERMANN, 1987; HÖVERMANN and HÖVERMANN, 1991; DERBYSHIRE et al., 1998), that the silt is deflated from accumulation areas of allochtonous sediments in the desert areas, for example dried lake basins or alluvial pediments and fanglomerates (bajadas). The latter extend for 20 to 40 km in the forelands of the mountains, cover a huge area of Asia and thus have to be regarded as one of the most important dust supplying environments (LIU et al., 1996; LEHMKUHL 1997a, b; 2000). Based on a comparative analysis of airfall dust and surface loess samples from the Chinese Loess Plateau, as well as on observations of a dust storm in 1993 in Gansu Province, northern China, DERBYSHIRE et al. (1998) concluded, that the surfaces of large piedmont alluvial fans of the Hexi Corridor, Gansu, were a main source for the Quaternary dust deposits in the western regions of the Loess Plateau.

Important dust and sand sources include the rivers, palaeo-lakes and palaeo-rivers of Central Asia. In addition, the Pleistocene pediments (fanglomerates) associated with widespread fluvial activity seem to constitute another main dust source. Besides lakes and rivers, dust is trapped by alpine meadows in the mountain areas. At present, the mean annual precipitation ranges from 200 mm to about 400 mm in these areas. However, in wetter parts, e.g. on the northern slopes of the Khangay, the formation of black montane soils dominates the dust accretion. This may also have been the case during the Middle Holocene climatic optimum in parts of the Tibetan Plateau. In Mongolia as well as in western Tibet, aeolian mantles consist of coarser sandy loess (LEHMKUHL, 1995, 1997b). In our view, the coarser sediments appear to consist of more local material, while the silts may represent the long distance dust flux. In addition, the vegetation cover in these areas is, or was, sparser than in those areas with a typical loess cover.

4.2 Distribution and dating of loess-like sediments in Mongolia

In the mountain areas of Western Mongolia aeolian, loess-like sediments and typical loess have been collected during five field season from 1994 to 1998 by the second author. Mantles of sandy silt can be found on top of the slopes in elevations between 2,000 to 2,700 m a.s.l.. These mantles are 0.5 to 1 m thick and they cover bedrock and solifluction debris. The material is mainly silt-sized with a variable content of fine sand up to 15 %. The widespread fanglomerates at the foothills of the mountains are predominantly covered with stone pavements as a wind stable surface. In some valleys toward the Uvs Nuur basin we found a higher content of sand in the covering layers. In addition, within the widespread pediments and fanglomerates of Central Asia as well as in tills and other moranic or fluvial sediments, silt could be found. In some sections with more than 1 m of this covering layer the loessic material below 1 m or so has 9 to 15 % carbon and up to 2 % of gypsum content. Especially in Mongolia the origin of calcium carbonate in these sediments is not clear, as there is no or only very limited limestone and gypsic rocks in the areas of Western Mongolia. The main source regions for carbonate are the lake basins. LEHMKUHL (1999b) suggests that during Glacial periods of the Pleistocene the calcium of silicate rocks was released by frost-weathering in the mountains. It was transported downhill by river systems and then concentrated in the closed basins where it was stored as biogenic $CaCO_3$. At the end of glacial periods during intervals with arid climatic conditions this carbonate together with silicatic silt became deflated and deposited in the aeolian mantles. These mantles may have been repeatedly eroded and re-deposited during the climatic cycles of the Pleistocene (Fig. 7).

The internal basin of the Zezeg Nuur (Fig. 1, No. 3) provides an example for the local transport and deposition of silt-sized particles. On the eastern bank of the Zezeg Nuur a small Holocene dune field is accumulated - but the content of sand sized particles is very low.

In the Uvs Nuur Basin we focus especially on the Turgen Kharkhiraa Mountains. On the different Pleistocene terraces the aeolian mantles can be distinguished by their weathering characteristics (GRUNERT et al. 2000, LEHMKUHL 1999a). The two lowermost terrace systems in different parts of the Mongolian Altai (T1, T2) are assigned to the last and penultimate Glaciations. The relation to the main ice margins and the weathering characteristics of the overlying stratum supports this preliminary stratigraphy (LEHMKUHL 1999a). On top of the first terrace a Holocene soil with minor laminated white calcrete on the underside of stones can be observed in horizons about 30 to 40cm below the surface. Palaeosoils and buried humic horizons date to the mid and younger Holocene. Strongly weathered aeolian sediments including minor calcrete and brownish clay skins often associated with stony horizons cover the second, and higher terrace. The alluvial fans and fanglomerates in the Uvs Nuur basin can also be separated by the degree of weathering of the calcretes (GRUNERT et al. 2000, LEHMKUHL 2000).

Sedimentological analyses show a differentiation of the aeolian mantles in the Turgen-Kharkhiraa Mts. in two main groups (Fig. 6). Below 2,200 m a.s.l. the grain size distribution maximum is represented by the sand fraction (see 5 curves in Fig. 6 above), and above 2,200 m a.s.l. more silty sediments dominate (see 6 curves in Fig. 6 below). Sandy aeolian sediments also dominate in the eastern Uvs Nuur Basin (Fig. 1, s. 1) and in the catchment area of the Orchon (Fig. 1, s. 4). However, the group of silty sediments can be compared with typical sandy loess from the Tibetan Plateau (LEHMKUHL, 1995, 1997a, b).

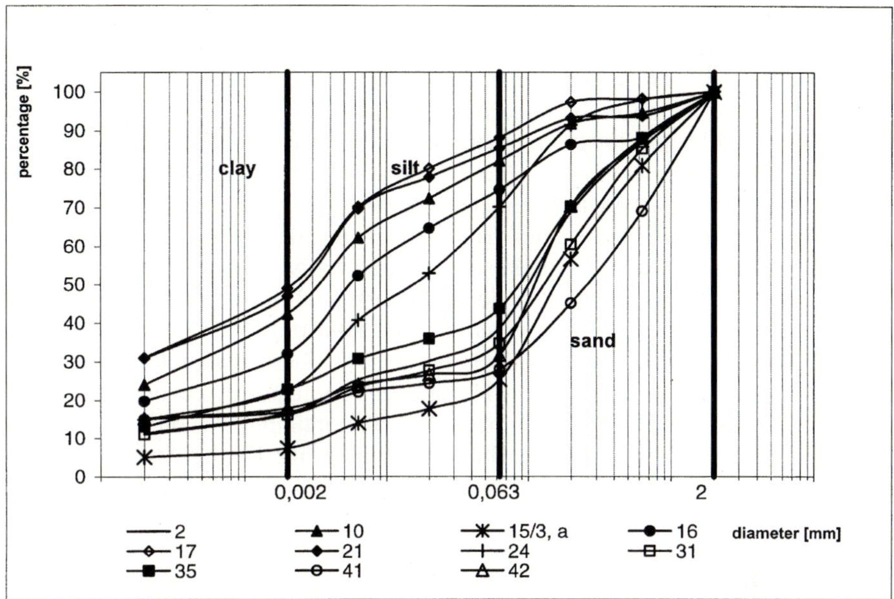

Fig. 6. Grain size distribution and sites of selected samples of covering sediments in the Turgen-Kharkhiraa Mts. (LEHMKUHL).

OSL-samples at the base of the aeolian cover sediments were taken at several places to provide a chronological framework for the Pleistocene in this mountain system. First luminescence data provided by J. REES-JONES, E. RHODES (Oxford) and A. LANG (Bonn, both personal communications) suggest that loess accumulation in these continental areas was intensified during the Interstadial periods, the Glacial periods and the Early Holocene. Two sections on the northern slope of the Turgen Kharkhiraa Mountains in the catchment area of the Khöndlön Gol and a fill of an ice wedge cast at the eastern slope of the Turgen Mountains provide Interstadial data (29.3 ±2.8; 37.4 ±3.7 ka, and 37.1 ±3.3 ka). Overlying strata on terraces of the Kharkhiraa Gol, and in the internal basin of the Huh Nuur in-between these two main rivers, date to the Holocene (5.6 to 8.4 ka). Further unpublished luminescence data from the southern slope of the Turgen Kharkhiraa mountains and from other areas of the Mongol Altai (LEHMKUHL et al., 2000) collected by the second author provide OSL data clustering in the Interstadial of the Last Glaciation, just after the LGM or Late Glacial to Early Holocene. More details and additional sections will be published in a forthcoming paper.

In addition, for aeolian and colluvial sediments that cover fluvial and glaciofluvial terraces in the central part of Khangay (Fig. 1, No. 4) LEHMKUHL and LANG (2001) provide Holocene TL and OSL ages. An age of 21 ka is obtained for a sand deposit overlaying the terrace, which is related to the Last Glacial ice margin. Lacustrine sediments from higher beach lines in the Valley of the Gobi Lakes (Fig. 1, No. 5) provide evidence for a slightly more humid period around 1.5 ka, and a larger extent of the lakes in the Early Holocene at about 8.5 ka, which can also be found in other areas of Central Asia. However, remnants of lacustrine sediments buried by alluvial gravel, and indicating a huge palaeo-lake in the basin of the Orog Nuur, date to the early stage of the Last Glaciation period around 70 ka (LEHMKUHL and LANG, 2001). In other areas of Western Mongolia there are actually no data available confirming the existence of last Interglacial loess. This is obviously the result of the widespread erosion during the Glacial stages.

In addition, luminescence dates from aeolian mantles on the Tibetan Plateau provide evidence for dust accumulation since the Early Holocene above 3,500 to 4,000 m a.s.l., and Pleistocene (Glacial and Interstadial) loess accumulation below this elevation in most sections, respectively (LEHMKUHL, 1995; LEHMKUHL et al., 2000). According to HOFMANN (1993) the zone of loess accumulation migrated into the modern arid forelands of the Helan Shan Mountains twice: once before the LGM and then again during the Holocene climatic optimum. According to ZHANG et al. (1994) loess accumulation on the north flank of the Kunlun Shan Mountains increased during the Holocene. ROST (1997) reported an intensified loess accumulation in the Qinling Shan just after the LGM (18 ka).

However, based on the morphostratigraphy and all other informations available, the genesis of aeolian, loess-like sediments can be sketched in four different phases (LEHMKUHL 1999b, Fig. 7).

(1) During glacial stages intensified weathering in the mountains produced large amounts of debris, sand, and silt. The material is transported by rivers towards the internal basins of Central Asia and accumulated in large alluvial fans. Finally, it will be transformed into lacustrine sediments. Calciumcarbonate ($CaCO_3$) is precipitated and concentrated in the lake basins.

(2) At the end of the Glacial stages, when the climatic conditions turned towards higher aridity, the lakes shrunk and the clay, silt and sand-sized particles can be eroded easily. Whereas the sand is transported by higher wind speed and accumulated not far in the main wind direction, the silt-sized particles can be eroded easily and represent the major source for the long distance transport toward the Chinese Loess Plateau and even the Pacific Ocean (NILSON and LEHMKUHL 2001). However, an unknown part of this silt-sized particle transport remain within the area and is accumulated in specific dust traps, e.g. the uppermost part of slopes and incorporated in solifluction layers and in the pediments accumulated in the mountain front.

(3) During Late-glacial periods and at the early Holocene when the climate turns to warmer and wetter conditions vegetated surfaces of a semiarid environment provide suitable surfaces for the trapping of dust. As seen in several sections, these mantles are silt-sized with a different content of fine sand (0.063-0.2 mm in diameter). The original content of gypsum and calcium carbonate at the base of some sections indicates the aeolian origin of these sediments.

(4) In the warmer Interstadial and/or Interglacial periods (e.g. Isotope stage 3 and 1 of the Holocene) soil development in the mountains and basins occur. Especially in the basins the soil development stopped the movement of dune sand.

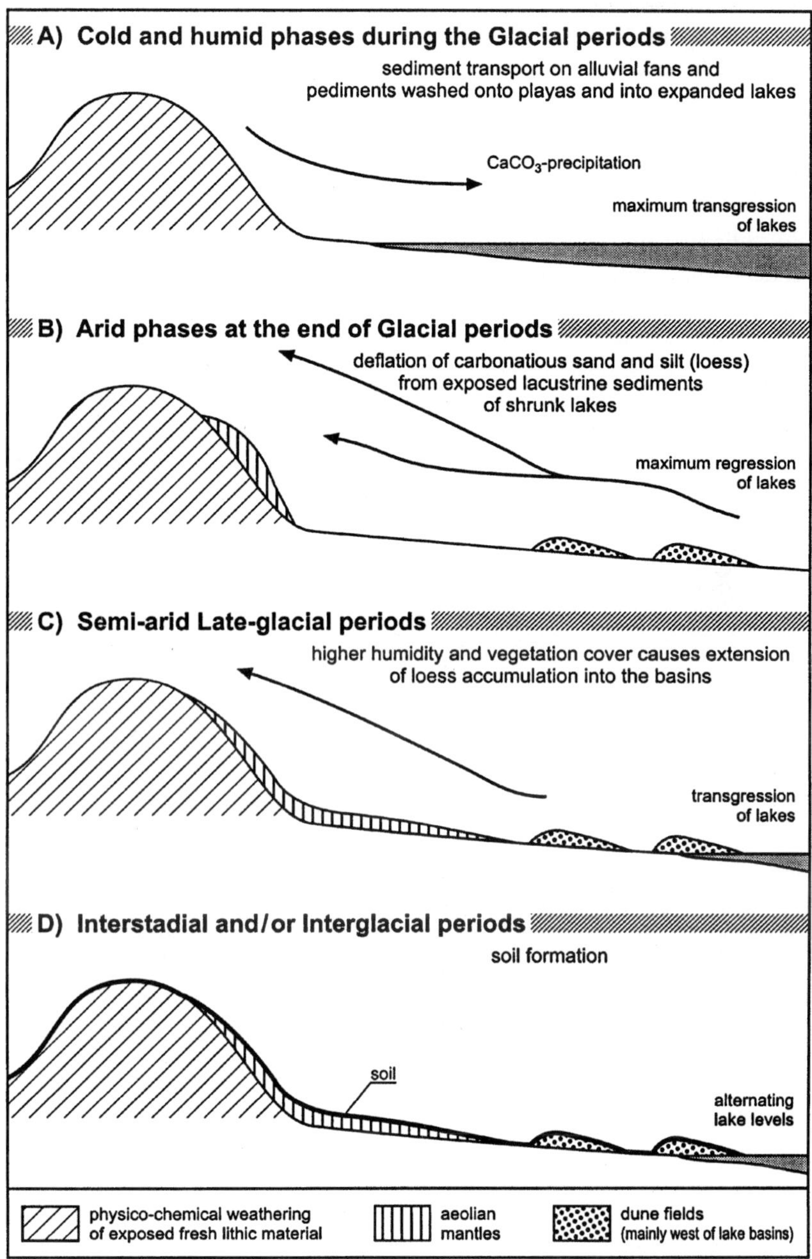

Fig. 7. Model of horizontal and vertical sediment transport in the basin and range area of Western Mongolia during the Pleistocene and Holocene depending on temperature (glacial and periglacial activity), aridity (aeolian transport and accumulation), and humidity (lake transgression and soil formation). Modified from LEHMKUHL (1999b).

However, in the Late Holocene the change towards drier climatic conditions caused shrinking lake levels in Central Asia and Mongolia. In this period, the dried out lacustrine surfaces have become a new dust source.

5 Conclusions

Aeolian sediments are very common in all parts of semi-arid Western Mongolia independent from altitude. They can be found in the large basins (about 1000 m a.s.l.) as well as on the flancs of high mountains up to 3000 m a.s.l. independent from slope aspect. Indeed, they can be divided into three groups: sand in the basins (Fig. 3), fine sand on the slopes of the foothills and lower mountains (Fig. 6, bottom), and silt above an altitude of about 2000 m a.s.l. (Fig. 6, top). The position of these belts may have changed vertically due to the climate fluctuations during the Pleistocene and Holocene ranging between arid and semi-arid conditions. More humidity could stop the aeolian activity completely because of the formation of soils everywhere. Periods like these are documented more spectacular by lacustrine sediments high above the present-day lake levels indicating a huge extension of former lakes.

The interaction between fluvial sediments, lake bottom sediments and aeolian sand can be well demonstrated by the example of Uvs Nuur Basin. It is possible to reconstruct a horizontal sediment cycle, which could have existed since the beginning of the Pleistocene. Indeed, fluvial erosion in the watershed of Tesijn Gol is estimated to have been very effective during the past, and the amount of sediment being transported towards the Uvs Nuur Basin must have been very high. Despite the effectivity of the postulated sediment cycle a continuous sedimentation can be supposed the thickness of which, unfortunately, is unknown. So, it will be difficult to calculate the sediment budget of the basin for modelling purposes.

Whereas the sand transport in the basin is more or less a horizontal sediment cycle (Fig. 5), the silt-sized particles or loess-like sediments are transported in a horizontal cycle and also in a vertical cycle, respectively (Fig. 7). The latter one is also supporting the long-distance transport and, therefore, this is the only sediment transport leaving the basins without outlet in the interior of Asia.

Acknowledgements

We are grateful to the German Research Foundation for the financial support of our projects. We wish to express our thanks to Dr. Dordschgotow, the vice-director of the Mongolian Academy of Science, and his secretary Tschimgee for the excellent cooperation during several years, and to our Mongolian partners during the expeditions in the far West of their country: Dr. O. BATKHISHIG, Dr. D. Dash and others. We also include our German assistants, J. BRAUNSDORF, Dr. M. KLEIN, Dr. M. KLINGE, and E. NILSON, and, last but not least, the speaker of the

German-Mongolian research group, Prof. Dr. U .TRETER. In addition, we would like to thank Prof. Dr. B. Meyer for soil analysis and several discussions.

References

AN, Z., WU, X., LU, Y., ZHANG, D., SUN, X., DONG, G. and WANG, S. (1991): Paleoenvironmental changes of China during the last 18,000 years. In: Liu Tungsheng (ed.): Quaternary geology and environment in China, 228-236.

BATKHISHIG, O. and LEHMKUHL, F. (1999): Soils of the Harhiraa Turgen mountains. - In: Ministry of Nature and Environment, WWF, Uvs Aimag and Lake Uvs Basin SPAs Administrations (Eds.): Global Change and Uvs Nuur – Sustainable development of the Altai-Sayan Ecoregion and transboundary nature conservation issues. International Conference, 6.-10.8.1999:140-155.

BERG, L.S. (1958): Die geographischen Zonen der Sowjetunion.-437 pp., Teubner-Leipzig.

DASH, D. (1999): Landscapes of sand massifs and the problem of nature conservation for Ikh Nuur Hollow. Dissertation Ulaanbaatar [in Mongolian with Russian abstract].

DASH, D. and TUMURBAATAR, E. (2000): Spatial differentiations of the Buurug Del Els landscape. - Berliner Geowiss. Abh., A, **205**, 9-13.

DERBYSHIRE, E., MENG, X. and KEMP, R.A. (1998): Provenance, transport and characteristics of modern aeolian dust in western Gansu Province, China, and interpretation of the Quaternary loess record. - Journal of Arid Environments, **39**, 497-516.

DEVJATKIN, E.V. and MURZAEV, E.M. (1989): The late Cenozoic of Mongolia - The Mongolian Altai. Nauka, 42-47. [in Russian].

DING, Z.L., RUTTER, N., HAN, J.T. and LIU, T.S. (1992): A coupled environmental system formed at about 2.5 Ma in East Asia. - Palaeogeography, Palaeoclimatology, Palaeoecology, **94**, 223-242.

DODONOV, A.E. (1991): Loess of Central Asia. - GeoJournal, **24**, 185-194.

DORDSCHGOTOV, D. (1992): Soils of Mongolia. Genesis, systematics, geography, resources and land use. Faculty of Soil Sciences; Moskau [in Russian].

DOROFEYUK, N.I. and TARASOV, P.E. (1998): Vegetation and lake levels in Northern Mongolia in the last 12500 years as indicated by data of pollen and diatom analyses. - Stratigraphy and Geological Correlation, **6 (1)**, 70-83.

FANG, J.Q. (1991): Lake evolution during the past 30,000 years in China, and its implications for environmental change. - Quaternary Research, **36,** 27-60.

FENG, Z.D., CHEN, F.H., TANG, L.Y. and KANG, J.C. (1998): East Asian monsoon climates and Gobi dynamics in marine isotope stages 4 and 3. - Catena, **33**, 29-46.

FENG, Z.D. (2001): Gobi dynamics in the Northern Mongolian Plateau during the past 20,000+ yr: preliminary results. - Quaternary International **76/77**: 77-83.

FLORENSOV, N.A., and KORZHNEV, S.S. (1982): Geomorphology of Mongolian People Republic. - Joined Sovjet-Mongolian scientific research geological expeditions. - Transactions, Vol. **28**. Moscow. [in Russian].

FRENZEL, B. (1994): Zur Paläoklimatologie der letzten Eiszeit auf dem tibetischen Plateau. - Göttinger Geogr. Abh., **95**, 115-142.

GRUNERT, J. (2000): Palaeoclimatic implications of dunes in the Uvs Nuur Basin, Western Mongolia. - Berliner Geowiss. Abh., **A, 205,** 2-8.

GRUNERT, J. and KLEIN, M. (1998): Binnendünen im nördlichen Zentralasien (Uws Nuur, westliche Mongolei). - Berliner Geogr. Abh., **63**, 45-66.

GRUNERT, J., KLEIN, M., STUMBÖCK, M. and DASH, D. (1999): Bodenentwicklung auf Altdünen im Uvs Nuur Becken. - Die Erde **130**, 97-115.

GRUNERT, J., LEHMKUHL, F. and WALTHER, M. (2000): Paleoclimatic evolution of the Uvs Nuur Basin and adjacent areas (Western Mongolia). - Quaternary International **65/66**: 171-192.

HAASE, G. (1978) Struktur und Gliederung der Pedosphäre in der regionischen Dimension. - Beiträge zur Geographie, Supplementbd., **29**/3. (Ed: Akad. d. Wiss. der DDR). Berlin, 250pp.

HILBIG, W., BASTIAN, O., JÄGER, E.J. and BUJAN-ORSICH, Ch. (1999): Die Vegetation des Uvs-nuur-Beckens (Uvs-Aimak, Nordwestmongolei). - Feddes Repertorium, **110**, 569-625.

HOFMANN, J. (1993): Geomorphologische Untersuchungen zur jungquartären Klimageschichte des Helan Shan und seines westlichen Vorlandes (Autonomes Gebiet Innere Mongolei/VR China).- Berliner Geographische Abhandlungen **57**. Berlin, 187pp.

HOFMANN, J. (1999): Geoökologische Untersuchungen der Gewässer im Südosten der Badain Jaran Wüste (Aut. Region Inner Mongolei/VR China) – Status und spätquartäre Gewässerentwicklung. – Berliner Geogr. Abh. **64**: 1-247.

HOVAN, S.A., REA, D.K., PISIAS, N.G. and SHACKLETON, N.J. (1989): A direct link between the China loess and marine δ18O records: aeolian flux to the north Pacific. - Nature, **340**, 296-298.

HÖVERMANN, J. (1987): Morphogenetic regions in Northeast Xizang (Tibet). In: Hövermann, J. and Wang W. (Eds.), Reports of the Qinghai-Xizang (Tibet) Plateau. Science Press, Beijing, pp. 112-139.

HÖVERMANN, J. and HÖVERMANN, E. (1991): Pleistocene and Holocene geomorphological features between the Kunlun Mountains and the Taklimakan Desert. - Die Erde, Ergänzungsheft, **6**, 51-72.

JÄKEL, D. (1995): Die Wüsten Chinas. Aufschlussreiche Zeugen globaler Klimaschwankungen. - Naturwiss. Rundschau, **10**, 365-373.

JÄKEL, D. (1996): The Badain Jaran Desert: Its origin and development. - Geowissenschaften **14**, 272-274.

KLEIN, M. (2000): Die hydrologische Wegsamkeit von Binnendünen des Uws Nuur Bekkens, nördliches Zentralasien. - Jenaer Geographische Schr., **9**, 23-37.

KLEIN, M. (2001): Binnendünen im nördlichen Zentralasien (Uws Nuur Becken, nordwestliche Mongolei). - Mainzer Geographische Studien, **47**, 182p.

KLINGE, M. (2001):Glazialgeomorphologische Untersuchungen im Mongolischen Altai als Beitrag zur jungquartären Landschafts- und Klimageschichte der Westmongolei. – Aachener Geographische Arbeiten 35. Aachen, 125p.

KOWALKOWSKI, A. and STARKEL, L. (1984): Altitudinal belts of geomorphic processes in the South Khangai Mts. (Mongolia). - Studia geomorphologica Carpatho-Balcanica, **18**, 95-115.

LEHMKUHL, F. (1995): Geomorphologische Untersuchungen zum Klima des Holozäns und Jungpleistozäns Osttibets. - Göttinger Geogr. Abh., **102**, 1-184.

LEHMKUHL, F. (1997a): Der Naturraum Zentral- und Hochasiens. - Geogr. Rundschau, **49**, 300-306.

LEHMKUHL, F. (1997b): The spatial distribution of loess and loess-like sediments in the mountain areas of Central and High Asia.-Z.Geomorph. N.F., Suppl.-Bd., **111**, 97-116.

LEHMKUHL, F. (1998): Extent and spatial distribution of Pleistocene glaciations in Eastern Tibet. - Quaternary International, **45/46**, 123-134.

LEHMKUHL, F. (1999a): Rezente und jungpleistozäne Formungs- und Prozeßregionen im Turgen-Kharkhiraa, Mongolischer Altai. - Die Erde, **130**, 151-172.

LEHMKUHL, F. (1999b): Cycles of loess formation during different periods of the younger Pleistocene in Central Asia. – In: DERBYSHIRE, E. (Ed.): Loessfest ′99, Extended Abstracts: 144-145.

LEHMKUHL, F. (2000): Alluvial fans and pediments in Western Mongolia and their implications for neotectonic events and climatic change. – Berliner Geowissenschaftliche Abhandlungen, Reihe A, **205**: 14-21.

LEHMKUHL, F. and HASELEIN, F. (2000): Quaternary paleoenvironmental change on the Tibetan Plateau and adjacent areas (Western China and Western Mongolia). - Quaternary International, **65-66**, 121-145.

LEHMKUHL, F. and KLINGE, M. (2000): Bodentemperaturmessungen aus dem Mongolischen Altai als Indikatoren für periglaziale Geomorphodynamik in hochkontinentalen Gebirgsräumen. - Zeitschr. f. Geomorphologie, N.F., **44**: 75-102.

LEHMKUHL, F., KLINGE, M., REES-JONES, J. and RHODES, E.J. (2000): First luminescence dates for Late Quaternary aeolina sedimentation in Central and Eastern Tibet. - Quaternary International, **68-71**: 117-132.

LEHMKUHL, F. and LANG, A. (2001): Geomorphological investigations and luminescence dating in the southern part of the Khangay and the Valley of the Gobi Lakes (Central Mongolia). - Journal of Quaternary Sciences **16**: 69-87.

LEHMKUHL, F., SCHLÜTZ, F., BECKERT, C. and KLINGE, M. (1998): Zur jungpleistozänen und holozänen Klimageschichte des Turgen-Charichira, Mongolischer Altai. - Jenaer Geographische Manuskripte **19**, 43-44.

LIU, T. XITAO, Z. JIAMAO, H. and HONGHAN, Z. (eds.) (1985): Loess and environment. Beijing, 251pp.

LIU, T, SHOUXIN, Z. and JIAOMAO, H. (1986): Stratigraphy and palaeoenvironmental changes in the loess of Central China. - Quaternary Science Reviews, **5,** 489-495.

LIU, T., DING, M. and DERBYSHIRE, E. (1996): Gravel deposits on the margins of the Qinghai-Xizang Plateau and their environmental significance. - Palaeogeography, Palaeoclimatology, Palaeoecology, **120.** 159-170.

MURZAEV, E.M. (1954): Die Mongolische Volksrepublik. Physisch-geographische Beschreibung. Gotha.

NAUMANN, S. and WALTHER, M. (2000): Mittelholozäne Seespiegelschwankungen des Bayan Nuur (Nordwestmongolei). - Marburger Geogr. Schriften, **135**: 15-27. Marburg.

NAUMANN, S. (1999): Spät- und postglaziale Landschaftsentwicklung im Bayan Nuur Seebecken (Nordwestmongolei). - Die Erde, 130, 117-130.

NILSON, E. (1998): Jungpleistozäne Staubflüsse in Asien und den zirkumasiatischen Ozeanen - Interpretation eines komplexen Klimasignals. - Diplomarbeit, Universität Bonn.

NILSON, E. and LEHMKUHL, F. (2001): Interpreting temporal patterns in the Late Quaternary dust flux from Asia to the North Pacific. - Quaternary International **76/77**: 67-76.

OPP, C. (1991): Erste Ergebnisse bodenphysikalischer, bodenchemischer und landschaftsökologischer Untersuchungen in der Mongolei. - Mitteilungen d. Deutschen Bodenkundlichen Gesellschaft 661, 197-200.

PACHUR, H.J., WÜNNEMANN, B. and ZHANG, H. (1995): Lake evolution in the Tengger Desert, Northwestern China, during the last 40,000 years. - Quaternary Research, **44**, 171-180.

PÉCSI, M. (1990): Loess is not just the accumulation of dust. - Quaternary International, **7/8,** 1-21.

PYE, K. (1996): The nature, origin and accumulation of loess. - Quaternary Sciences Reviews, **14**, 653-667.

PYE, K. and ZHOU, L.-P. (1989): Late Pleistocene and Holocene aeolian dust deposition in north China and the northwest Pacific Ocean. - Palaeogeography, Palaeoclimatology, Palaeoecology, **73**, 11-23.

QIN, B. and YU, G. (1998): Implications of lake level variations at 6 ka and 18 ka in mainland Asia. - Global and Planetary Change, **18**, 59-72.

RICHTER, H., BARTHEL, H. and HAASE, G. (1961): Klimamorphologische Höhenstufen des zentralen Changai in der Mongolischen Volksrepublik. - Geographische Berichte, **20/21**, 162-168.

ROST, K.T. (1997): Observations on distribution and age of loess-like sediments in the high-mountain ranges of central China. - Zeitschrift f. Geomorphologie, N.F., Suppl.-Bd., **111**, 117-129.

SMALLEY, I.J. (1995): Making the material: The formation of silt-sized primary mineral particles for loess deposits. - Quaternary Science Reviews, **14**, 645-651.

SMALLEY, I.J. and KRINSLEY, D.H. (1978): Loess deposits associated with deserts. - Catena, **5**, 58-66.

TARASOV, P.E. et al. (1996): Lake status records from the FSU, Database documentation Version 2. IGBP PAGES/World Data Center-A for Paleoclimatology Data Contributions Series 96-032. NOAA/NGDC Paleoclimatology Program, Boulder CO, USA.

TARASOV, P.E. and HARRISON, S.P. (1998): Lake status from the former Soviet Union and Mongolia: a continental-scale synthesis. - Paläoklimaforschung / Palaeoclimate Research, **25**, 115-130.

TSOAR, H. and PYE , K. (1987): Dust transport and the question of desert loess formation. - Sedimentology, **34**, 139-153.

WALTHER, M. (1998): Paläoklimatische Untersuchungen zur jungpleistozänen Landschaftsentwicklung im Changai-Bergland und in der nördlichen Gobi (Mongolei). - Petermanns Geographische Mitteilungen, **142**, 205-215.

WALTHER, M. (1999): Befunde zur Seespiegel- und Klimaentwicklung in der Nordwest-Mongolei. - Die Erde **130**, 131-150

WALTHER, M. and NAUMANN, S. (1997): Beobachtungen zur Fußflächenbildung im ariden bis semiariden Bereich der West- und Südmongolei (Nördliches Zentralasien). - Stuttgarter Geogr. Studien, **126,** 154-171.

WRIGHT, J., SMITH, B. and WHALLEY, B. (1998): Mechanisms of loess-sized quartz silt production and their relative effectiveness: laboratory simulations. Geomorphology, **23**, 15-34.

WÜNNEMANN, B. (1999): Untersuchungen zur Paläohydrographie der Endseen in der Badain Jaran- und Tengger Wüste, Innere Mongolei, Nordwest-China. - Habil. Schr. am FB Geowiss. der FU Berlin.

WÜNNEMANN, B., PACHUR, H.-J., LI, J. and ZHANG, H. (1998): Chronologie der pleistozänen und holozänen Seespiegelschwankungen des Gaxun Nur / Sogo Nur und Baijian Hu, Innere Mongolei, NW-China. Petermanns Geographische Mitteilungen, **142**, 191-206.

ZHANG, X., AN, Z., CHEN, T. and ZHANG, G. (1994): Late Quaternary records of the atmospheric input of eolian dust to the center of the Chinese loess plateau. - Quaternary Research, **41**, 35-43.

Ostracod ecology of alluvial loess deposits in an eastern Tian Shan palaeo-lake (NW China)

Mischke, S.[1], Hofmann, J.[2] & Schudack, M.E.[1]

[1]Institute of Palaeontology, Free University of Berlin,
Malteserstr. 74-100, 12249 Berlin, Germany
palaeont@zedat.fu-berlin.de

[2]Institute of Geographical Sciences, see above

Abstract

An alluvial loess sequence of 105 m thickness was studied in a steep mountainous desert valley in the eastern Tian Shan. Although no very clear remnants of a natural dam have yet been found, we conclude that probably a huge landslide downstream of the sequence resulted in a dam rising at least 105 m above the valley floor. Seven ostracod species including *Ilyocypris bradyi*, *Eucypris lilljeborgi*, *Darwinula stevensoni* and *Limnocythere inopinata* were found in the sediments and facilitated the reconstruction of the lake history. After the valley was filled suddenly, swamp conditions prevailed, leading to a fluctuating shallow lake and swamp environment at first and the formation of a deeper and more stable lake afterwards. Shallow water and swamp conditions again prevailed, succeeded by a more stable and deeper lake, until conditions changed once more to a shallow lake with a dense subaquatic vegetation. A last deeper lake period was followed by a second period of a shallow, densely vegetated lake. Freshwater conditions prevailed more or less during the entire lake evolution.

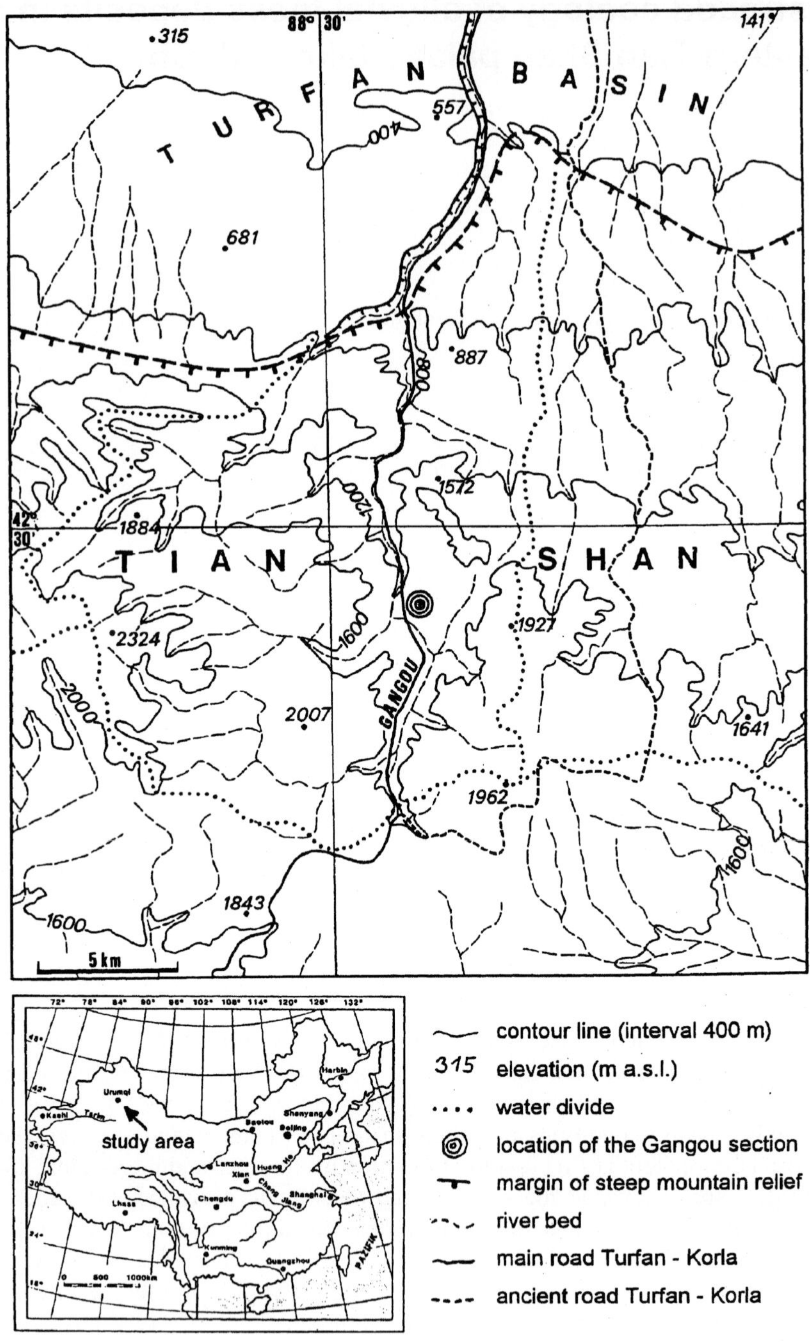

Fig. 1. Location of the study area in eastern Tian Shan

1 Introduction

In late summer of 1999 a 2-year cooperative palaeoecology project between the Institute of Palaeontology (Free University of Berlin) and the Department of Geography (Lanzhou University) commenced with a 3-week field survey in the drainage system of Bosten Hu (Xinjiang/P.R. China). During the initial reconnaissance the authors had the opportunity to investigate also the southwestern part of the Turfan Basin and the adjoining parts of Tian Shan.

The area of the present investigation (42°20' to 42°40'N, 88°20' to 88°40'E) is located at the easternmost edge of Tian Shan (Heavenly Mountains), one of the world's great mountain ranges. The eastern branch of the Tian Shan is also the southern boundary of the Turfan Basin, an intermontane foreland basin. The central part of the Turfan Basin, known as the Turfan depression, has a minimum altitude of -154 m a.s.l., and is the second lowest exposed land surface on Earth. The extremely recent elevation of the mountain range in the study area is certainly due to the Cenozoic India-Asia collision, being evidence of substantial uplift of the range. Owing to this uplift, the relief is characterized by steep slopes, sharp crest lines as well as deeply incised valleys and gorges. The Gangou valley (Fig. 1), which is one tributary to the Turfan Basin, comprises a catchment area of approx. 300 square kilometres with the highest point at 2007 m a.s.l. and the lowest at the valley outlet being 550 m a.s.l. Therefore the general valley gradient is in the range of 6.5 m/100m. At an elevation of 1240 m a.s.l. a section of fine-grained material with a total thickness of 105 m was discovered and studied by the

Fig. 2. Alluvial loess profile of the Gangou section

authors. The exact location of this section is 42°28’,615 N; 88°31,879 E, 100 m east of the main Turfan-Korla road. At this site fine-grained sediments are exposed at the steep valley flanks covering an area of less than 1 square kilometer (Fig. 2). The former extent of these fine-grained deposits is not known, but remnants are clearly visible on the opposite slope. Most of the year the Gangou is a dry valley with episodic runoff after rainfall events, therefore the arid conditions favour the conservation of these sediments.

The existence of a thick sequence of fine-grained sediments in such a strange situation within a valley of steep relief with usually bare rock on the slopes certainly gives rise to some questions. Why do sediments of low-energy conditions occur in a mountainous desert relief which has in the most recent past been characterised principally by erosion? After the shift to a depositional environment, what caused the termination of these conditions and the onset of recent processes? What sort of environment prevailed at the time of deposition of the sediments in question? Are there signs of environmental fluctuations?

2 Materials and methods

The sequence was studied by detailed visual inspection of the sediments; in particular, lithology, grain size, colour, presence of macrofossils (e.g. mollusc shells) and sedimentary structures. Samples of about 200 to 550 g were taken at more or less equally spaced intervals over the section. Sampling procedure was focused on fine-grained layers to obtain material deposited in a low-energy environment which was expected to contain microfossils. Samples were disaggregated with water and sieved through a 100 µm mesh.

Ostracod valves were separated under a microscope, valves and fragments of determinable size were counted and absolute abundances calculated for a standardised sample size of 500 g. Relative abundances were not given because of general low numbers of ostracod valves and the high variability.

Hand specimen of different exposed crystalline rocks were sampled in the close vicinity of the section for petrographical analysis. Thin sections were prepared and investigated under a petrographic microscope.

3 Climatic and geological setting

The climatic conditions are characterized by an extreme continental desert climate with cold winters. The following data refer to the weather-station Turfan (42°56’N, 89°12’E, elevation 34.5 m a.s.l., observation period 1951-1980) approx. 70 km northeast of the Gangou section (DOMRÖS & PENG 1988). Annual rainfall is 16 mm, therefore the centre of the Turfan basin is the driest location known in China. The annual potential evaporation measured at Turfan is 2838 mm. The mean annual air temperature is 13.9°C with the lowest monthly mean temperatures in January (-9.5°C) and the highest in July (32.7°C). In winter minimum temperatures of -28.0°C and below occur, while the absolute maximum temperature is 47.6°C. Because of the lack of data for the eastern edge of Tian Shan the annual precipitation is not known for the study area. The rain shadow effect created by the Bogda Shan-Barkol Tagh ranges in the north indicates that the mountains at the southwestern border of the Turfan Basin receive only a small amount of rainfall (< 100 mm).

The fine-grained sediments under investigation were deposited within a steep valley incised into mainly slaty volcanic rocks of Palaeozoic age (Allen et al. 1993). The rocks exhibit a predominant N-S or WNW-ENE striking of schistosity and, in less foliated members, of joint planes. Most frequent rock types are intermediate or acid volcanics of different grain sizes and colours (middle to dark grey, some types slightly reddish or blueish), mostly with a considerable amount of quartz. Most volcanics are fine-grained. Coarse porphyroblastic structures are rare except for some volcanic dykes exhibiting floating structures of large hornblendes.

To the north of the narrowest part of the valley, large intrusions of granitoids form the steep cliffs of the mountains. These light grey intrusions have been penetrated and covered by dark volcanic rocks. Both rock units are also of Palaeozoic, most probably Carboniferous age. Large boulder streams have derived from the highly weathered granitoids, filling the valley from time to time.

4 Results

The investigated sequence with a thickness of 105 m is mainly composed of alternating mud and sand layers (Fig. 3). Single mud layers are mainly about 1 to 5 cm thick but can reach 20 cm at maximum. They are characterised by a light-brown colour, the frequent occurrence of plant molds and either an internal fine lamination or an absence of internal structures. Especially in the top of the section thicker mud beds of the same kind may be found. Mudcracks were not observed in the section.

Sand layers are predominantly 5 to 10 cm thick, fine-grained and normal graded. Rhizoliths cemented by carbonate or iron oxides are common throughout the sand beds of the whole sequence, whereas mud clasts of gravel size occur only occasionally. Coarse-grained sand layers or even gravel layers occur within the lower 20 m of the section and show a dark colour similar to that of the exposed crystalline bedrock in the vicinity of the sequence. Sometimes the sand beds are ripple-stratified or cross-bedded with current ripples or foresets indicating a north-easterly or northerly flow direction during deposition. However, internal structures in the sand beds are usually poorly defined or absent.

The upper half of the sequence is indicated by the predominance of thicker mud beds which are hard, whitish-grey and rich in carbonate. In contrast to internal structures, horizontal bedding planes of different layers are well defined, and their thickness does not change laterally.

Ostracods belonging to 7 species have been found in the sediments of the section. *Ilyocypris bradyi* (Sars, 1890) and *Eucypris lilljeborgi* (G.W. Müller, 1900) reach high abundances in some samples of the upper third of the sequence (stage IV and V, Fig.3) but occur in low specimen numbers also in stage II. *Darwinula stevensoni* (Brady & Robertson, 1870), *Limnocythere inopinata* (Baird, 1843) and *Candoniella* sp. generally show abundances not exceeding 40 specimens per sample. *D. stevensoni* and *L. inopinata* show a simultaneous occurrence and abundance, but in contrast to the former *L. inopinata* is not only present in stage III and IV but already in stage II with a few specimen. The other two species, *Candona neglecta* Sars, 1887 and *Cyprideis torosa* (Brady, 1868) , are present in one sample only with two valves each.

5 Discussion

The predominantly alternating mud and sand layers interrupted by coarser sand layers or occasionally gravels and the more continuous beds of either mud or sand are generally interpreted as alluvial loess deposits. The absence of mudcracks or signs of initial soil formation points to a more or less continuous existence of a waterbody, which was probably very shallow, as indicated by the frequent occurrence of rhizoliths and plant molds nearly throughout the whole of the section.

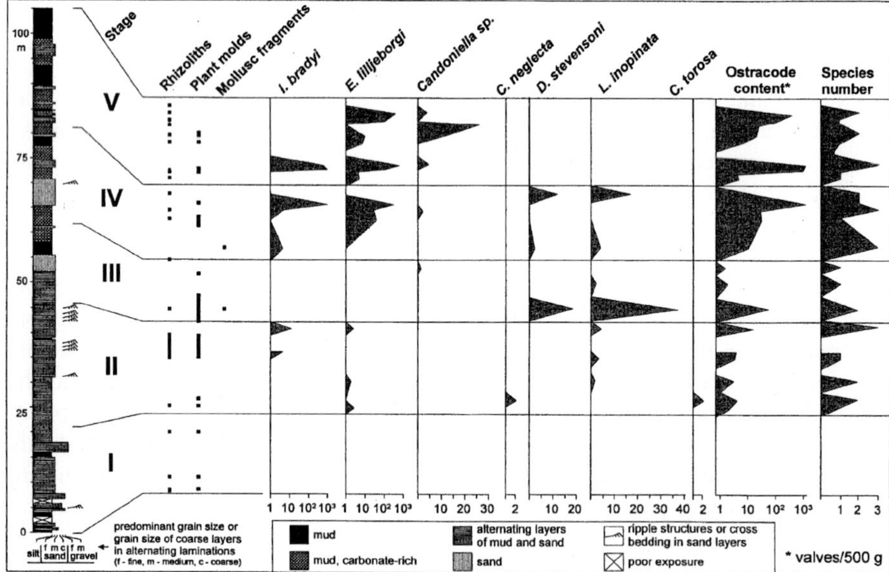

Fig. 3. Sediments and abundance of ostracod species

Although the waterbody was shallow, very stable thicknesses of sets and laterally steady layers show that it was probably large in area. So we assume that the valley was filled by a very shallow lake with plant growth at least at the margins. A lake like this suggests the presence of a natural dam which had filled the whole valley floor downstream of the sequence. From the thickness of the sequence we can reconstruct an elevation of at least 105 m for the lowest point of the dam above the recent valley floor. The almost continuous existence of plants throughout the lake´s evolution indicates not only that the lake was shallow but also that the dam was permeable to some degree and/or that water inflow and sediment input were in balance to fill the basin gradually without intermittent deep water periods. After the creation of the dam the newly formed reservoir was not filled by water up to the crest of the dam. Instead, the basin probably contained shallow water all the time and the lake level rose higher and higher as a response to the infilling of the basin. The lake evolution was terminated either after the lowest point of the dam was reached and subsequently washed away or by a collapse of the dam before. At present it is only possible to speculate about the origin and position of the dam. Owing to the steep relief of the mountainous desert gorge and the resulting strong erosion it is difficult to distinguish possible remnants of a former dam from common accumulations of large boulders within the valley floor downstream of the sequence. However, large boulder streams derived from highly weathered granitoids in the most narrow part of the valley only a few hundred meters to the north of the Gangou section suggest that the palaeo-dam had also evolved from these rocks in the close vicinity of the lake sediments.

Further information about the lake history may be derived from the ostracod ecology of present taxa.

C. neglecta is present in the sequence only as fragments which show distinct signs of transport on the surface (Fig.4), whereas *C. torosa* is untypically thin-shelled and present only as juveniles.

Because both species have been found in one sample only, *C. neglecta* obviously underwent postmortem transport and because of the weakly calcified appearance of *C. torosa* both taxa are not considered to be representative of the section. The ecological requirements of the other taxa are summarised to reconstruct the evolution of the palaeo-lake.

I. bradyi is a holarctic cold water species, which is typical of fresh and shallow waters although it can withstand slightly saline conditions. It is commonly found either in permanent and intermittent streams (Delorme 1970) or in temporary ponds or springs. It is a weak swimmer and can be found predominantly burrowing or crawling among aquatic plants and organic debris (De Deckker 1979). It is one of the species most tolerant of low oxygen concentrations (below 3 mg/l, Mezquita et al. 1999). The specimen collected as a result of this study show "marginal ripplets" on the distal parts of the inner lamellae of left valves (Plate 1). The ripplets are slightly less well defined than those of specimen used by Janz (1994) as a valuable tool to distinguish the most abundant recent species of the genus *Ilyocypris,* but they are well in agreement with the description by Van Harten (1979) who recognized this shell character for the first time.

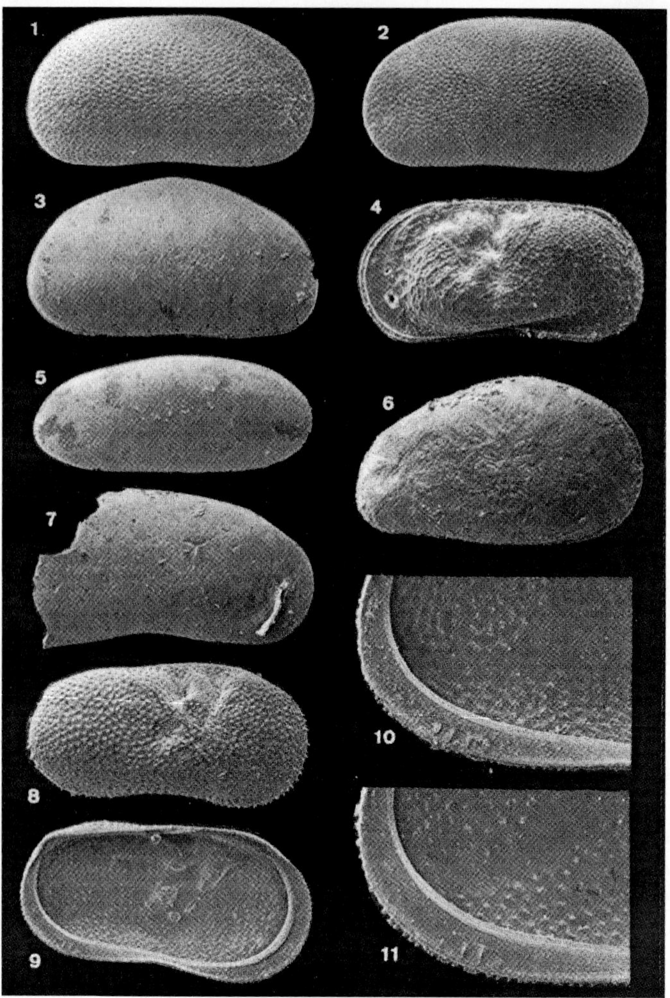

Fig. 4. Scanning electron micrographs of ostracod species from the Gangou section *Candoniella* sp.: 1 – RV, lv, L: 630 µm, H: 340 µm, 2 – LV, lv, L: 640 µm, H: 350 µm, *Eucypris lilljeborgi* (G.W. Müller, 1900): 3 – RV, lv, L: 1240 µm, H: 700 µm, *Limnocythere inopinata* (Baird, 1843): 4 – LV (juvenile), lv, L: 480 µm, H: 250 µm, *Darwinula stevensoni* (Brady & Robertson, 1870): 5, LV (juvenile), lv, L: 430 µm, H: 190 µm, *Cyprideis torosa* (Brady, 1868): 6 – RV (weakly calcified juvenile), lv, L: 540 µm, H: 320 µm, *Candona neglecta* Sars, 1887: RV (fragment, marks of postmortem transport on surface, length: 930 µm), lv, *Ilyocypris bradyi* Sars, 1890: 8 – RV, lv, L: 820 µm, H: 430 µm, 9 – LV, iv, L: 790 µm, H: 430 µm, 10 – LV, iv, enlargement of posteroventral inner lamella showing "marginal ripplets" typical for the species, width of cutting: 330 µm, 11 - LV, iv, enlargement of figure 9, posteroventral inner margin showing "marginal ripplets" typical for the species, width of cutting: 300 µm.
(Abbr.: RV – right valve, LV – left valve, lv – lateral view, iv – internal view, L – length, H – height)

E. lilljeborgi lives in small, temporary waterbodies and ponds and has been recorded in Holocene as well as in recent sediments from Central to Southern and Eastern Europe over Iran to the Karakorum Mountains (Pietrzeniuk 1985).

Species of *Candoniella* such as *C. lactea* or *C. albicans* have been found in Quaternary deposits of the Kunlun Mountains (Li et al. 1997), the Tarim Basin (Sun et al. 1999), the Qaidam Basin (Chen & Bowler 1986), the Weihe Basin (Zhang Zonghu 1990) and the Zoige Basin of Sichuan Province (Wang & Zhu 1991) and are considered to be freshwater and oligohaline species.

D. stevensoni is a cosmopolitan species living typically in the shallow waters of lakes and ponds. It is a eurythermic freshwater species which can tolerate oligohaline and occasionally mesohaline salinities (Mazzini et al. 1999). According to Ranta (1979) this species seems to avoid flowing water. Although unable to swim and restricted to crawling at the sediment surface or within the sediment it requires well ventilated surface sediments (Grafenstein et al. 1999).

L. inopinata is very adaptable; it is common in fresh to polyhaline waters, polythermophilous and belongs to the phytal assemblage, although it has also been found in greater depths of lakes (Hiller 1972, Carbonel et al. 1988). It is widespread all over Northwestern China and the Tibetan Plateau and the dominant species in modern Qinghai Lake (Peng et al. 1998). Its stratigraphical range in the Qaidam Basin and in the Tarim Basin covers approximately 1 Ma up to the present (Yang et al. 1997, Sun et al. 1999).

The presence or absence of the ostracod taxa from the Gangou section and their abundances were considered to characterize 5 different stages of lake evolution (Fig. 3).

Stage I: The first stage is distinguished by the deposition of alternating layers of mud and sand in which the sand beds are the coarsest grained ones of the whole sequence, even containing gravel layers occasionally. Rhizoliths and plant molds occur, but are not frequent. Ostracods are totally absent in stage I. This stage is interpreted as belonging to the initial phase of lake basin formation. The beginning of this phase is not recorded in the investigated sequence because we presume several meters of sediment between the underlying bedrock and the base of our sequence. During stage I the valley floor had already been levelled by alluvial deposits, which were subsequently covered by very shallow water. No proper river channel deposits or structures have been found. Single layers of coarse sands or gravels were spread over the floor of the shallow waterbody as a result of the short distance to the entering river and/or of floods following extraordinary precipitation events. The swamp environment was probably still too capricious for the establishment of an ostracod fauna. One can argue that there was no lake during stage I because the precipitation/evaporation ratio (P/E) was too small and/or the dam which was probably made of unweathered rocky material was too permeable. By contrast, a lake could have existed during this stage in a lower position in the valley where lake sediments are no longer preserved on the steep valley slopes.

Stage II: Almost exclusively alternating mud and sand layers were deposited during stage II with sand layers frequently showing structures of flowing water. The depth of the waterbody probably fluctuated resulting in either swamp or shallow lake conditions with a reduced or extended distance to the river mouth respectively. Shallow and unstable lake conditions are indicated by the first appearance of only a few ostracods within the sequence. The fragmented and juvenile valves of *C. neglecta* and *C. torosa* were most likely transported from upstream during very shallow water conditions as indicated by the simultaneous absence of *E. lilljeborgi.*

Stage III: First pronounced stable lake conditions are marked by the simultaneous appearance of *D. stevensoni* and *L. inopinata* about 49 m above the base of the sequence. The simultaneous appearance of both taxa seems to be very typical of modern shallow, stagnant lake waters in Northwestern China. The water depth was still shallow, i.e. not exceeding a few meters. A thick sand bed starting about 52 m above the base coincides with the termination of the first stable lake period and probably reflects a decrease of the water level. Following this episode the sediments of the sequence are predominantly fine-grained; mainly mud or carbonate-rich mud indicates an extended distance from the shore, a higher residence time, a lower river energy and/or filtering by a denser vegetation belt. The upper half of stage III shows a return to fluctuating conditions similar to those during stage II.

Stage IV: This stage may be divided into three substages with stable and deeper conditions in the lower part around 65 m above the base as indicated by the simultaneous appearance of *D. stevensoni* and *L. inopinata* and an intermediate substage with shallow water and a dense subaquatic vegetation marked by the occurrence of *I. bradyi*, *E. lilljeborgi* and *Candoniella* sp. as well as the absence of the former two species (around 75 m) and a third period with a return to yet again deeper and more stable conditions (around 79 m). Both sets of environmental conditions (deeper, more stable lake and shallow water with dense subaquatic vegetation) are distinguished quite clearly by the occurrence of either *D. stevensoni* and *L. inopinata* or *I. bradyi* and *E. lilljeborgi*. The absence of plant molds during the periods around 65 m and 79 m above the base and the presence of mollusc fragments during the former period are further evidence for deeper and more stable lake conditions.

Stage V: The complete absence of *D. stevensoni* and *L. inopinata* within stage V is taken as proof that deeper and more stable lake conditions, like those in stage IV, did not come into existence. Instead, a shallow lake with a dense vegetation on the bottom persisted for nearly the entire stage V. Short-term interruptions were attributed to episodes of a very shallow water depth and/or increasing river energy.

On the whole, deeper and more stable lake conditions existed during periods of stages III and IV. It is very doubtful whether present P/E conditions could enable the formation of a permanent lake at the position of the Gangou section with a catchment area of only approx. 100 square kilometers. Therefore we expect a distinctly higher P/E ratio for the appropriate periods at least, although we have not yet been able to connect this to the forcing mechanism.

The fine-grained material which has filled the palaeo-lake itself certainly does not derive from the Palaeozoic rocks close to the lake, but is more likely to have been transported into the area over longer distances by the wind. It is also uncertain whether an aeolian loess blanket of considerable depth formerly covered the slopes of the mountains or whether simultaneous dust input and wash into the palaeo-lake occurred.

Up to now we have no evidence for the age of the section from dated material. Very rough estimations might be made by considering the stratigraphical range of the ostracod taxa which are well known from the large basins of Northwestern China. Because *L. inopinata* is not known from the sediments of the Qaidam Basin and the Tarim Basin before 0.97 Ma and, in addition, *C. neglecta* appeared not before approximately 1.2 Ma (Yang et al. 1997) the age of the section may be roughly classified as Middle or Upper Quaternary. With respect to the very soft erodible material still present at some places we assume a Weichselian or even Holocene age. Further investigations are needed to shed more light on these questions.

6 Conclusions

An alluvial loess sequence of 105 m thickness was studied in a steep mountainous desert relief in the eastern ranges of Chinese Tian Shan. The formation of the mountainous basin was most probably the result of a huge landslide which dammed the valley up to an elevation of at least 105 m above the valley floor. The alluvial loess deposits of the investigated sequence contain 7 species of Ostracoda such as *I. bradyi*, *E. lilljeborgi*, *D. stevensoni* and *L. inopinata.* The former two are typically found in temporary ponds and springs or even in permanent and intermittent streams (*I. bradyi*) whereas the latter are very common in lakes. Based on ecological requirements of modern ostracods like these and on sedimentological characteristics the alluvial loess sequence has been divided into 5 stages. The swamp environment of the first stage was succeeded by a fluctuating shallow lake and swamp environment. A more stable and deeper lake came into being for the first time within the third stage, but was followed again by shallow water and swamp conditions. During the fourth stage deeper and more stable conditions prevailed until the environment changed to a shallow lake with a dense subaquatic vegetation. A last period of deeper water and more stable conditions occurred once again before a shallow densely vegetated lake prevailed during nearly the entire last stage.

These results enable us to reconstruct the palaeohydrological as well as the palaeo-environmental conditions of the lake. The link to the palaeoclimate will be the next step of further research.

Acknowledgements

The expedition program was financed by the Deutsche Forschungsgemeinschaft (DFG, Schu 694/10-2). S.M. is grateful to the Deutscher Akademischer Austauschdienst (DAAD) for a Graduate Fellowship (Grant No. D/97/23142). Special thanks are due to Prof. Zhang Hucai from Lanzhou University for administrative assistance.

References

Allen, M.B.; Windley, B.F.; Zhang, Chi & Guo, Jinghui 1993: Evolution of the Turfan Basin, Chinese Central Asia. - Tectonics, 12, No. 4: 889-896.

Carbonel, P., Colin, J.-P., Danielopol, D.L., Löffler, H. & Neustrueva, I. 1988: Palaeoecology of limnic ostracods: A review of some major topics. - Palaeogeogr., Palaeoclimatol., Palaeoecol., 62: 413-461.

Chen Kezao & Bowler, J. M. 1986: Late Pleistocene Evolution of Salt Lakes in the Qaidam Basin, Qinghai Province, China. - Palaeogeogr., Palaeoclimatol., Palaeoecol., 54: 87-104.

De Deckker, P. 1979: Middle Pleistocene ostracod fauna of the West Runton Freshwater Bed, Norfolk. – Palaeontology, 22: 293-316.

Delorme, L.D. 1970: Freshwater ostracode of Canada. Part IV. Families Ilyocyprididae, Notodromadidae, Darwinulidae, Cytherideidae, and Entocytheridae. – Canadian Journal of Zoology, 48: 1251-1259.

Domrös, M. & Peng Gongbing 1988: The Climate of China. Springer, Berlin.

Grafenstein, U.v., Erlenkeuser, H. & Trimborn, P. 1999: Oxygen and carbon isotopes in modern fresh-water ostracod valves: assessing vital offsets and autoecological effects of interest for palaeoclimate studies. - Palaeogeogr., Palaeoclimatol., Palaeoecol., 148: 133-152.

Hendrix, M.S.; Trevor, T.A. & Graham, S. 1994: Late Oligocene-early Miocene unroofing in the Chinese Tian Shan: An early effect of the India-Asia collision. - Geology 22: 487-490.

Hiller, D. 1972: Untersuchungen zur Biologie und zur Ökologie limnischer Ostracoden aus der Umgebung von Hamburg. - Arch. Hydrobiol., Suppl., 40 (4): 400-497.

Janz, H. 1994: Zur Bedeutung des Schalenmerkmals ‚Marginalrippen' der Gattung Ilyocypris (Ostracoda, Crustacea). - Stuttgarter Beitr. Naturk., Ser. B, 206: 1-19.

Li Yuanfang, Li Bingyuan, Wang Guo, Li Shijie & Zhu Zhaoyu 1997: Ostracoda and ist environmental significance at the ancient Tianshuihai Lake of the West Kunlun. - Journal of Lake Science, 9: 223-230.

Mazzini, I., Anadon, P., Barbieri, M., Castorina, F., Ferreli, L., Gliozzi, E., Mola, M. & Vittori, E. 1999: Late Quaternary sea-level changes along the Tyrrhenian coast near Orbetello (Tuscany, central Italy): palaeoenvironmental reconstruction using ostracods. – Marine Micropalaeontology, 37: 289-311.

Mezquita, F., Tapia, G. & Roca, J.R. 1999: Ostracoda from springs on the eastern Iberian Peninsula: ecology, biogeography and palaeolimnological implications. - Palaeogeogr., Palaeoclimatol., Palaeoecol., 148: 65-85.

Peng Jinlan, Zhang Hucai & Ma Yuzhen 1998: Late Pleistocene limnic ostracods and their environmental significance in the Tengger desert, Northwest China. - Acta Micropalaeontologica Sinica, 15: 22-30.

Pietrzeniuk, E. 1985: Ostrakoden aus dem holozänen Travertin von Weimar. – Z. geol. Wiss., 13: 207-233.

Ranta, E. 1979: Population biology of Darwinula stevensoni (Crustacea, Ostracoda) in an oligotrophic lake, - Ann. Zool. Fennici, 16. 28-35.

Wang Jingzhe & Zhu Dajin 1991: Quaternary Ostracoda from Hongyuan and Zoige Regions, Sichuan. - Acta Micropalaeontologica Sinica, 8: 111-119.

Sun Zhencheng, Feng Xiaojie, Li Dongming, Yang Fan, Qu Yonghong & Wang Hongjiang 1999: Cenocoic Ostracoda and palaeoenvironments of the northeastern Tarim Basin, western China. - Palaeogeogr., Palaeoclimatol., Palaeoecol., 148: 37-50.

Van Harten, D. 1979: Some new shell characters to diagnose the species of the *Ilyocypris gibba – biplicata – bradyi* group and their ecological significance. – In: Krstic, N. (ed.): Taxonomy, biostratigraphy and distribution of ostracods. Serbian Geol. Soc., Beograd: 71-75.

Yang Fan, Sun Zhencheng, Ma Zhiqiang & Zhang Yonghua 1997: Quaternary ostracode zones and magnetotratigraphic profiles in the Qaidam Basin. - Acta Micropalaeontologica Sinica, 14: 378-390.

Zhang Zonghu 1990: Explanatory notes of the Quaternary geologic map of the People's Republic of China and adjacent sea area. China Cartogr. Publ. House, Beijing.

Critical comments on the interpretation and publication of ^{14}C, TL/OSL and $^{230}Th/U$ dates and on the problem of teleconnections between global climatic processes

Dieter Jäkel

Institute of Geographical Sciences
Free University of Berlin
Malteserstr. 74-100, Haus H
D-12249 Berlin

Abstract

Modern research in the earth sciences includes the use of physical dating methods to clarify stratigraphical issues. Without including such results, no paper is likely to be accepted for publication by a well-reputed journal. This does not always seem justified, especially in those cases where the paper contains a complete relative chronology as a case study for a distinct region. On the other hand, papers presenting methodologically unreliable age data are accepted and later serve as the basis for teleconnections concerning global climatic patterns. Most of these age determinations are conducted by commercial laboratories without taking account of the relative chronology of the strata concerned. Their reliability has to be evaluated by the author who – on the other hand – receives no information about possible difficulties of sample preparation or about the complexity of the method and the measurements. Data obtained by routine laboratory techniques may not reflect the degree of reliability necessary for correct interpretation and global correlation. This may be the case in radiocarbon dating: for example, if the apatite and collagen contents of bones are not determined separately, or the reservoir effect is ignored when analysing water and biogenic carbonates. This situation is particularly unfortunate because detailed studies of the problem are available.

1 Introduction

Towards the end of the 2001 IGCP Meeting in Rauischholzhausen (26-28 January) the future activities of the project group were discussed. The question was raised whether it makes sense to submit TL/OSL samples taken by individual members of the group to different national or international laboratories for dating even though there is no satisfactory way of establishing their reliability. TL experts present at the discussion were of the opinion that such dating results are often merely "laboratory-specific" and of no scientific value. There is something too random about dates obtained by "routine procedures", since as a rule there is no exchange of information between the submitter and the analyst. For this reason, commercial laboratories were able to provide results at a relatively low price. Since physical and chemical laboratory procedures have improved over the years, it seems superfluous to discuss them. However, problems occur in those cases where the physical parameters of a sample have to be estimated rather than measured. The requisite double or multiple analysis is omitted because of the costs involved. The consensus was that it would be better in future to have samples analysed in laboratories that release the results only if they stand up to critical examination in all respects. However, this presupposes that these laboratories receive financial support to offset the high labour and material costs involved. The discussion ended with the conclusion that more funding has to be obtained for methodological improvements in order to meet the latest standards of age determination in the natural sciences.

Having been concerned about these issues since I started to submit samples for dating (Jäkel, 1971; Geyh & Jäkel, 1974), I welcome this debate and the resulting proposals as long overdue. Despite past experience, too many colleagues have resigned themselves to the present dubious situation. The aim of this paper is therefore to recall earlier - but still valid – findings by M.A. Geyh at the ^{14}C dating laboratory of the *Niedersächsisches Landesamt für Bodenforschung* in Hannover. He has always been concerned with the issue of correct dating and has achieved methodological improvements in sometimes very complex procedures, always in agreement with the submitters, who reap the benefits in the end. This is especially true of the ^{14}C dating conducted in his laboratory with respect to humic acids, the apatite and collagen content of bones, and the reservoir effect in water and biogenic carbonates.

2 Reliability of ^{14}C dates

Full cooperation between field workers or submitters and geochronologists always involves the common goal of obtaining dates that are as correct as possible. In this context, the problem of contamination by humic acids arose during the 1960s and 1970s. In general, contaminated samples yielded too-young ages, especially in the case of older material. This problem was addressed in detail in the 1950s and 1960s already by DeVries (1958) and I. Olsson (1979). Dating of samples from the Tibesti Mountains (central Sahara) revealed that the residual material documented the time of deposition, the acids a later infiltration by a wet climate phase (Jäkel 1971, p. 31, 44/45; Geyh & Jäkel, 1974, p. 107ff; Jäkel & Geyh, 1982). This was possible because the exact stratigraphic location of the sample material was known. If this is not the case, and contaminated material has to be dated, experience has shown that the results are probably unreliable. The inverse effect due to contamination by humic acids was found in the case of glacigenic samples from the Berlin area (Böse, 1979, p. 28, 34).

Here the samples in question were peats and peat material from Weichselian moraine deposits. The dates - based on samples that had not been pretreated - indicated a probable early ice advance of the Nordic ice sheet into the Berlin area about 30,000 years ago. Since there was no other evidence of this, Böse arranged for multiple analysis of the samples in Hannover. Despite good sampling conditions and initially appropriate material, the samples turned out to be unsuitable for dating purposes. Very different values were obtained for the humic acids: either they agreed with the ages of the NaOH-insoluble fraction, were too young, or as much as 18,000 years too old (Böse, p. 25, 28). In the case of the too-high dates Geyh pointed out that infiltrating humic acids sometimes coalify rapidly and no longer dissolve in sodium hydroxide solution, therefore cannot be separated from the remaining fraction of the sample. At the time, Geyh foresaw the necessity of determining the conditions under which such coalification may occur and of recognizing and excluding them during sampling and laboratory procedures (Böse, p. 34). Since humic acids enter the samples and/or sampled horizons via infiltration and groundwater the possibility of varying contamination by reservoir effects within a very small area should be taken into account.

The influence of the reservoir effect has been known for the past 30 years and more. Geyh, Merkt & Müller (1971) published a paper of fundamental importance on this subject. Geyh proposed correcting the values for the Tibesti region by 1800 to 2800 years (Jäkel & Geyh, 1982, p. 145) and adjusted the interpretation of his chronostratigraphic histograms accordingly. In the meantime these methodological considerations have received further attention and new results have been presented that help to clarify the issues (Hofmann & Geyh, 1998; Arp, G., Hofmann, J. & Reitner, J., 1998; Hofmann 1999). These publications supply convincing evidence of the effort required to obtain correct dates. Because the authors conducted their analyses on material from biogenic mats that had started to form in 1927 owing to an earthquake-induced displacement of a lake floor, they obtained varying ^{14}C reservoir ages: 1080 ± 155 (Hv 21925) for microbial mats, an average of 2140 ± 160 for sandy muds, and 3220 ± 225 years for the carbonate fraction (Hofmann & Geyh, 1998, p. 96). Recent results from the Atacama suggest that the reservoir effect also changes over time.

A glance at recently published radiocarbon dates shows that the majority lack any comments as to reliability. It no longer seems of much interest how a sample was processed in the laboratory, hence no information is given about the issues described above. So even expert analysts can no longer assess the reliability of the given dates which stand uncommented, without even approximately meeting the requirements formulated by Geyh & Schleicher (1990: 21/22). They therefore lack both the precision generally associated with the method and the comparability essential for stringent scientific discussion. The possible consequences will be described in Section 5 below.

3 The uranium/thorium method

Since the ^{14}C dating method has a limit of ~50,000 years BP, other techniques had to be developed to date older material. An important method for Quaternary research is the $^{230}Th/U$ method, which yields correct dates up to 500,000 years old (Geyh & Schleicher, 1990, p. 222). Many physical and chemical criteria are involved and if even one of these is not fulfilled the date may no longer be representative (Geyh & Schleicher, 1990, p. 213). This shows that the preparation and analysis of samples is extremely labour-intensive and that it should largely be left to the expert analyst to interpret the data, the role of the field worker being to supply information about the sites and settings of the samples. The following examples show how problematical this issue may be: during an expedition to the Autonomous Province of Inner Mongolia in China in 1988 we sampled lacustrine sediments at the once huge palaeolake of Gashu-Sogo Nor at the cliff bordering Sogo Nor, and yardangs to the west of Wentugaole. In Hannover, these samples were dated using both the ^{14}C and the U/Th methods (Table 1).

The comments on the U/Th ages by Henning and Geyh dated 1 December 1989 state that "sample Uh 571 contained about twice as much Th as U and therefore required a relatively large age correction for detrital Th-230. A presumed correction factor of 1 yields a U/Th age of 159 ± 9 ka, and a correction factor of 2 - which is altogether conceivable - gives a U/Th age of 120 ka. The true sample age could therefore point to marine isotope stage 5, assuming the sample was definitely not exposed to surface water over a long period." – "In the case of the calcrete, "Sogonor" of Uh sample 573 the Th/U ratios are much more favourable, so the age correction is slight here and minimally affected by the choice of correction factor. Nevertheless, the above-mentioned reservations apply if the sample was possibly an 'open system' "(author's transl.).. It should not be forgotten that an absolute desert climate was assumed then and that the commentary would be different today.

A comparison of our ^{14}C dates with U/Th ages shows that the ^{14}C dates are in the finite range. They are therefore an unsuitable basis for a definitive statement.

But the U/Th ages are extremely uncertain also and therefore unsuitable for accurate dating of the sediments. For this reason we have refrained from interpreting and publishing the dates in the context of climate change.

In 1994, in the Wentugaole region at an elevation of 1000 m, we discovered two 2-3 cm thick lake carbonate horizons at 12.50 and 13.20 m in a 30 m high section characterized by a fivefold alternation of aeolian and lacustrine sediments and thought we had found better dating material. Accompanied by M. A. Geyh, we took samples in 1995, which were processed under the lab numbers Uh 1250 and 1254. After extremely detailed analysis of the samples, Geyh commented that "both layers have proved to be 54300 ± 3400 years old, both formed fairly closed systems, yet their uranium contents differ substantially, hence the water column will have been different." (author's transl.) In conversation, Geyh reacted with reserved scepticism to my opinion that the dates would fit in very well with a high lake level of Palaeolake Gashu Sogo Nur during the ice age. Therefore, in this case too, we consider it appropriate not to use these results for a definitive age dating, despite better analysis values.

In the southeast of the Badain Jaran Desert, depressions between the up-to-400 m high dunes contain more than 100 lakes, which were sampled and analysed by J. Hofmann (1999). In 1988 we took two samples of calcretes at a vertical interval of 4 m from conformable strata in a tectonically displaced palaeoprofile. They were analysed using both the ^{14}C and the U/Th methods (Hv 16095, Uh 661; Hv 16097, Uh 662, see Table 1), with the following comments by Henning and Geyh (dated 1 June 1990): "With a high Th-230/Th-232 activity ratio of 25, sample Uh 661 yielded a very reliable U/Th model age that is hardly affected by detrital correction. With a value of 2.9, sample U/Th 662 had a much lower Th-230/Th-232 activity ratio, which is why the corrected U/Th age is strongly dependent on the choice of correction factor: approx. 40 ka when fo = 1.0 is, approx. 32 ka when fo = 1.5." In this case too, even after the intervening methodological improvements it is assumed that the dates cannot be used without detrital correction.

The reliability of U/Th dates is strongly dependent on whether post-depositional mobilization of uranium can be excluded, i.e. whether systems were closed. Checks were made using ^{14}C analyses and finite ^{14}C dates were obtained in all cases. The ^{14}C age of sample Uh 661 - 21,000 years - is so low that an open system is likely. The true age probably lies in the 20 - 150 ka range.

At 14 ka, the difference between the ^{14}C and U/Th ages of sample Uh 662 is slight. Here too, the 'true' age probably lies between the two limits (25 ka and 40 ka)." (author's transl.). It is evident that a time range of 20 to 150 ka is not appropriate for a chronostratigraphic interpretation of sample Uh 661. Its age can probably be narrowed down by comparison with sample Uh 662. Hence we assume that the "true" age of sample Uh 661 is around 40 ka. That may sound somewhat unsubstantiated, which is why it should be pointed out again how senseless it is to publish "bare" dates without commenting on the reliability of the results.

Table 1. ^{14}C and U/Th dates of laminated carbonate samples of Palaeolake Gashu Sogo Nor and from the Badain Jaran Desert in China.

Sample No.	Site	Elevation a.s.l.	Hv No.	$\partial^{13}C$ $^{o}/_{oo}$	^{14}C pmc	conv. ^{14}C-age (carbonate) BP	Uh No.	Uranium content ppm	Thorium content ppm	U/Th age model fo=o	U/Th age corr. fo= 1.0
Jä 880922-3	Sogo Nor	920m	16085	+ 14.5	2.7 ± 0.1	29 010 ± 445	573	0.236 ± 0.003	0.166 ± 0.005	116 000 ± 4200	93 000 ± 3400
Jä 880929-2	Wentugaole	1000m	16088	+ 0.5	1.6 ± 01	33 285 ± 650	-	-	-	-	-
Jä 880929-3	Wentugaole	1020m	16089	+ 0.3	2.8 ± 01	28 430 ± 405	571	1.462 ± 0.023	2.726 ± 0.066	219 000 + 16 000 - 14 000	159 000 + 9000 - 8000
Jä 881006-4	Badain Jaran		16095	- 4.5	7.5 ± 0.2	20 800 ± 200	661	13.987 ± 0.333	1.696 ± 0.076	150 000 + 13 000 - 11 000	147 000 +12 000 -11 000
Jä 881006-7	Badain Jaran		16097	- 2.0	4.5 ± 0.1	24 900 ± 290	662	2.186 ± 0.050	1.115 ± 0.099	55 200 ± 5000	39 500 ±4300

Table 2. Results of TL dating of sands from Badain Jaran Desert and Palaeolake Gashu-Sogo Nor in China.

Lab No.	Field No.	U (ppm)	Th (ppm)	K (%)	Water levels	Plateau (^{O}C)	ED (Gy)	Dose rate (Gy/ka)	TL age (ka)
94086	Mi 940910-4	0.982	4.36	1.53	2	330-370	329 ± 20	2.39	138 ± 13
94087	Mi 940917-2	1.140	4.78	1.53	2	330-370	285 ± 15	2.47	115 ± 10
94088	Mi 940917-4	0.808	3.27	1.13	2	360-390	180 ± 20	1.83	98 ± 13
94089	Mi 940918-15	1.220	5.45	1.32	2	360-390	685 ± 50	2.31	296 ± 29
94090	Mi 940918-23	0.971	3.97	1.53	2	350-370	435 ± 30	2.34	186 ± 18
94091	Jä 940928-1	0.972	3.27	2.29	2	350-370	590 ± 40	3.10	190 ± 19
94092	Jä 940928-2	0.637	2.31	2.33	2	330-380	225 ± 20	2.99	75 ± 8.7

4 The thermoluminescence dating method

In Quaternary research TL and OSL dating techniques are gaining importance as a supplement to ^{14}C and U/Th methods, not least due to the fact that dates can be obtained from unconsolidated dune sand, fluvial sand and loess. Thus a previous gap in climatic interpretation is bridged, because geomorphological landforms such as fluvial and lake terraces, windblown sands, dunefields and loess deposits - which, especially in China, often belong to an ensemble or sequence of landforms - can now be fitted into a chronological setting as constituent elements of the landscape. For this reason we also took samples of windblown sand in underlying layers of the Heihe alluvial fan, the zibar sands in the Badain Jaran Desert (Mischke 1996), as well as from the first, diagenetically compacted, oldest generation of dune sand at the eastern edge of the Badain Jaran Desert. They were analysed in the TL/OSL Laboratory of the Geological Institute of the State Seismological Bureau in Beijing. The results are shown in Table 2.

In doubt about the accuracy of the dates, we asked Ludwig Zöller for his opinion. He came to the conclusion that because of methodological difficulties the dates should be interpreted as maximum ages, possibly too old by a factor of 2. However, the TL dates preclude a high- or lateglacial age (Dr. L. Zöller, pers. comm., 30 June 1996; Mischke, p. 67). The dates were used in Steffen Mischke's *Diplom* thesis, with diagrams showing their location in the profile complexes and commentaries. In our opinion, the dates – like the U/Th dates from this area, whose sediments represent the above-mentioned series - only suggest a high age and are not suitable for a detailed description of a climatic sequence. Therefore no further details will be given here to prevent possible speculation.

5 The problem of teleconnections

After describing some sources of error in age determination, we shall now return to the subject of processing published data. In many cases data are stored in computer files for later conversion into spatial and temporal distribution patterns. If the age data are highly accurate, a good computer program should yield a correct space-time analysis, but if erroneous data are processed the result is like an omelette whose flavour is spoilt by bad eggs or just one bad egg. Points of similar age are given that do not belong together at all. Since the problems and correction factors described in the original publications are not included, the program ignores them. The confidence interval is lost. In any case it would only disturb the calculations because the information is ambiguous. For this reason, teleconnections made on this basis are highly questionable. They discredit methods of age determination that allow us to compare contemporaneous features. Statistical constraints cause a well-researched field section to be suddenly shifted into a completely different time interval where field sampling parameters show that it does not belong. In our opinion this creates considerable problems, for such statistically erroneous procedures take reality *ad absurdum*. In the context of climate analyses of such global connections, major process changes - such as those from moist to dry or cold to warm - seem to occur at the same time but in reality, owing to regional peculiarities, they precede or lag behind the general trend. Hence, a spatial distribution pattern that is in reality complex and diverse is turned into something homogeneous. The result is worthless and it does not help when - as happened recently - authors are aware that data are erroneous and incomplete (Guo et al. 2000). Another completely unacceptable procedure in our view is to create time windows grouping data of specific time spans such as 3000 or 5000 years (12,000-9000 a BP or 15,000-10,000 a BP)),thus blurring boundaries, suppressing small-sized oscillations, and disguising true turning points. In comparison, frequency distribution diagrams like those developed by M. A. Geyh are substantially more informative because they also take account of the statistical error limits (Jäkel & Geyh, 1982, p. 143). Erroneous data that do not fit into the time setting are generally widely scattered and thus merely raise the baseline, but do not blur the peaks and gaps.

Here too, it should not be forgotten that the peaks and gaps are not a measure of the intensity of a fluctuation. They only show that at these times much or little material was found for age dating. However, fieldworkers tend to try and find as much material as possible at the transitions between formation processes, such as the start of a depositional phase (after incision of a river bed) or vice versa. Since material is available for depositional phases only, periods of erosion and deflation can only be reconstructed by sampling areas where the eroded material was redeposited. In morphological terms, this means that, for example, in a dryland area terrace deposits are eroded in the upper and middle reaches during a moist phase and redeposited again in the lower reaches and mountain forelands. Hence data are available for the start and the end of each depositional phase, but they only document the fact that this was a period of upheaval. Qualitative statements can only be made in the overall context of the sample setting. The detailed stratigraphy of the sediment columns in the profiles supplies further material for dating and complementary techniques such as particle-size, pollen and morphoscopic analysis. In this way, it is sometimes possible to identify erroneous data, as described above. Consistent local and regional chrono-histograms can then be constructed as a basis for teleconnections. If these conditions are not fulfilled, teleconnections are no more than fragmentary. Unfortunately such work often falls short of these requirements and should not be accepted for publication, even though it sometimes receives great publicity.

6 Summary

Discussion at the close of the 2001 IGCP meeting in Rauischholzhausen about TL dating prompted the author to write a paper about aspects of the reliability of ^{14}C, U/Th and TL dates. This seemed all the more urgent as such data often supply the basis for teleconnections. He therefore described some examples known to him from personal experience. Their dates of publication may make them appear obsolete, yet they have lost none of their fundamental value, as is confirmed by the fact that data before 1975 are also used. Attention is drawn to the danger of contamination by humic acids, which is largely a problem in the case of material more than 20,000 years old. Reservoir effects have to be considered when dating water and biogenic carbonates. A serious shortcoming is that recently only "bare data" have been published with no information about sample preparation or parameters indicating the reliability of the results. Table 1 shows ^{14}C and U/Th dates of the same sample in order to document how difficult it can often be to obtain the "true" sedimentation date of a sample. The TL dates in Table 2 are given with the same intention. As already stated in the descriptive texts, we explicitly emphasize again that the data in the tables are not suitable for purposes of comparison. Finally, attention is drawn to the problems that arise when teleconnections are not based on reliable dates or multiple analysis and to the fact that such comparative work should only be carried out by experts in geochronology. The author recommends setting up comprehensive data banks optimally containing all analyses of relevant parameters. He considers chrono-histograms to be a better means of representation than time slices. In any case, data used in teleconnections should be backed up by additional and complementary geoscientific methods. The fact that considerable deficits exist in this respect was the reason for this critical contribution.

7 Bibliography

Arp, G.; Hofmann, J. & Reitner, J. (1998): Microbial fabric formation in spring mounds („Microbiolites") of alkaline salt lakes, Badain Jaran sand sea, P.R. China. – Palaios 13, pp 581-592.

Böse, M. (1979): Die geomorphologische Entwicklung im westlichen Berlin nach neueren stratigraphischen Untersuchungen. – Berliner Geogr. Abh., 28, 51 p., Berlin.

De Vries, H.I. (1958): Variation in concentration of radiocarbon with time and location on earth. Kon. Ned. Akad. Wet. Proc. Ser. B 61: 94-102.

Geyh, M.A. & Jäkel, D. (1974.a): Spätpleistozäne und holozäne Klimageschichte der Sahara aufgrund zugänglicher ^{14}C-Daten. – Z. Geomorph., N.F., 18, pp. 82-98, Stuttgart-Berlin.

Geyh, M.A. & Jäkel, D. (1974, b): ^{14}C-Altersbestimmungen im Rahmen der Forschungsarbeiten der Außenstelle Barkai/Tibesti der Freien Universität Berlin – FU Berlin, Pressedienst Wissenschaft 5/74, pp. 107-117, Berlin.

Geyh, M.A.; Merkt, J. & Müller, H. (1971): Sediment-, Pollen- und Isotopenanalysen an jahreszeitlich geschichteten Ablagerungen im zentralen Teil des Schleinsees. – Archiv für Hydrobiologie, 69, pp. 366-399, Stuttgart.

Geyh, M.A. & Schleicher, H. (1990): Absolute Age Determination; Physical and Chemical Dating Methods and their Application. 503p., Springer-Verlag, Berlin-Heidelberg.

Guo, Z.; Petit-Maire, N. & Kröpelin, St. (2000) : Holocene non-orbital climatic events in present-day arid areas of northern Africa and China. – Global and Planetary Change, 26, pp. 97-103, Elsevier.

Hofmann, J. (1999): Geoökologische Untersuchungen der Gewässer im Südosten der Badain Jaran Wüste (Aut. Region Innere Mongolei/VR China) – Status und spätquartäre Gewässerentwicklung.- Berliner Geogr. Abh., 64, 247 p., Berlin.

Hofmann, J. & Geyh, M.A. (1998): Untersuchungen zum ^{14}C-Reservoireffekt an rezenten und fossilen lakustrinen Sedimenten aus dem Südosten der Badain Jaran Wüste (Innere Mongolei/VR China). – Berliner Geogr. Abh., 63, pp. 83-98, Berlin).

Jäkel, D. (1971): Erosion und Akkumulation im Enneri Bardagué-Arayé des Tibesti-Gebirges (zentrale Sahara) während des Pleistozäns und Holozäns. – Berliner Geogr. Abh., 10, 55 p., Berlin.

Jäkel, D. & Geyh, M.A. (1982): ^{14}C-Daten aus dem Gebiet der Sahara, hervorgegangen aus Arbeiten der Forschungsstation Bardai und des Niedersächsischen Landesamtes für Bodenforschung in Hannover. - Berliner Geogr. Abh., 32, pp. 143-165, Berlin.

Mischke, St. (1996): Sedimentologische Untersuchungen zur Landschaftsgenese der nordwestlichen Badain Jaran Shamo (Innere Mongolei/VR China): Diplomarbeit in der Fachrichtung Physische Geographie am Institut für Geogr. Wissenschaften der FU Berlin. 118 p. unveröffentlicht, Berlin.

Olsson, I.U. (1979): A warning against radiocarbon dating of samples containing little carbon. Boreas 8: 203-207.

INDEX

I List of Locations

II Subject Index

(very common items like „desert", „climatic change", or „palaeoecology" are not included in the index)

Druck: Strauss Offsetdruck, Mörlenbach
Verarbeitung: Schäffer, Grünstadt